Interviewing for Education and Social Science Research

INTERVIEWING FOR EDUCATION AND SOCIAL SCIENCE RESEARCH
THE GATEWAY APPROACH

Carolyn Lunsford Mears

palgrave
macmillan

INTERVIEWING FOR EDUCATION AND SOCIAL SCIENCE RESEARCH
Copyright © Carolyn Lunsford Mears, 2009.

First published in hardcover in 2009 by
PALGRAVE MACMILLAN®
in the United States—a division of St. Martin's Press LLC,
175 Fifth Avenue, New York, NY 10010.

Where this book is distributed in the UK, Europe and the rest
of the world, this is by Palgrave Macmillan, a division of Macmillan
Publishers Limited, registered in England, company number 785998,
of Houndmills, Basingstoke, Hampshire RG21 6XS.

Palgrave Macmillan is the global academic imprint of the above
companies and has companies and representatives throughout the
world.

Palgrave® and Macmillan® are registered trademarks in the
United States, the United Kingdom, Europe and other countries.

ISBN: 978–1–137–50793–8

Library of Congress Cataloging-in-Publication Data
Mears, Carolyn L.
    Interviewing for education and social science research:
    The gateway approach/Carolyn Lunsford Mears.
        p. cm.
    Includes bibliographical references and index.
    ISBN 978-0–230–61237-7
    1. Education—Research—United States. 2. Social sciences—
    Research—United States. 3. Interviewing—oral history—
    educational connoisseurship. I. Title.
    LB1028.M368 2009
    370.7'232—dc22   2009002649

A catalogue record of the book is available from the British Library.

Design by Integra Software Services

First PALGRAVE MACMILLAN paperback edition: April 2015

10 9 8 7 6 5 4 3 2 1

I dedicate this volume to my beloved A-B-C blocks,
Austin, Brian, and Connally,
the foundation of my life's joy and soul's delight.

CONTENTS

FOREWORD

Writers don't write from experience, although many are hesitant to admit that they don't. . . . If you wrote from experience, you'd get maybe one book, maybe three poems. Writers write from empathy.

Nikki Giovanni

Nikki Giovanni's insistence on the centrality of *empathy* as a precondition to writing exemplifies a necessary shift away from the separation implied by objectivity, and a movement toward connection if one is to embrace postmodernist ideals in empirical research. The very soul of postmodernism remains a devotion to radical reappraisal of modern assumptions about culture, identity, history, or language, as well as attention to developing empirical research methods that break down unproductive distances/barriers/boundaries between researcher and research participant. Instead of stripping emotion, chaos, and unruliness from research, as had been the practice in the modern era, postmodern approaches seek to engage the complexity of lived experience and to close the distance between researcher and researched. As such, contemporary efforts in empirical research, especially research that depends primarily on qualitative data, seek to open to examination issues formerly difficult to examine and to reinject phenomena long dismissed by modernists as ephemeral, such as emotions, passions, and illogical thought or behavior. This book and the journey that led to its creation contribute to this effort in a variety of ways, and among its most significant accomplishments is its contribution to providing a transparent example of how to develop a piece of research grounded in a researcher's empathy for her research participants. Carolyn Mears evokes the need for empathy when she sets the tone of her book, writing: "Successful interviewing *requires* attention to another's point of view" (p. 4, emphasis added).

Some years ago, when our now-grown children were in elementary school, Carolyn and I shared carpool duties, driving from our west-side suburban community to a school near downtown Denver. Rather than a close social relationship, ours was a relationship of convenience, and when our children went their separate ways, so did we. In 1987,

Carolyn moved to the Columbine area, and in late 1999, I moved from Colorado to assume a position at Wayne State University, in Detroit. Then, 20 years after carpooling with Carolyn, while I served as a reader on AERA's Qualitative Methods Special Interest Group Dissertation Award Committee, I opened my mail to find her dissertation presented for consideration. In the years between the 1980s and 2005, we had shared one central event: the Columbine attack. She was a parent of a student in the building that fateful day; I was more removed but had supervised a student teacher who had taken a job there and survived the attack.

We were both deeply affected by the event and, in our own ways, felt that the media circus simply could not grasp such a complex situation. Our dismay and frustration led us to undertake different research activities. Using anthropological studies of peer groups in high school, I wrote critically about "standardized" schools and brought into question whether schools like Columbine offered an education worth having to all students. Mine was sometimes an unwelcome critique inside the Columbine community. Carolyn used her insider standing to examine parents' experiences after the Columbine attack. Since AERA's recognition for her dissertation's contribution to qualitative methodology, we have become closer as colleagues, and I admire both her contributions to advancing the frontiers of research methodologies that depend on qualitative data and her efforts to develop a text that could be used in research methods courses, especially by students whose data set will come from interviews. Thus, I have had a front-row seat watching her ideas that first came to my notice as the methods chapter of a dissertation expand into a book consonant with postmodernist ideals, a book not couched in so much theoretical language that it will be inaccessible to graduate students.

A paucity of texts exist that cover the territory encompassed in *Interviewing for Education and Social Science Research: The Gateway Approach.* I have for several years taught the qualitative research methods sequence in Wayne State's College of Education. Like all professors, I struggle to keep reading material content high, but the overall cost to students low. Part of my dilemma is that so much of the methodological material is written for professors of method courses, not for novices, or is either too particular and doesn't cover enough territory, or too generic and provides only superficial coverage.

Carolyn's book strikes a middle ground. It not only clarifies and makes it easier to appreciate seemingly arcane ideas about research methods, but also pulls these together from sources that are spread across a wide range of academic disciplines. For instance, the work of Elliot Eisner is widely appreciated by scholars interested in aesthetics and research methods that unpack such a complex terrain. Some ideas in Eisner's work, *connoisseurship* for instance, provide a vantage point consistent with postmodernism and have broad applicability, but have not percolated as widely as they might.

Also, this book introduces *The Gateway Approach*. This comprehensive method ties together the ethical responsibilities of the researcher with decisions about data collection, fieldwork, and analysis. Though having broad applicability, it extends the field by making explicit how ethical practices, study design, study implementation, analysis, data presentation strategies, and research rigor should inform one another. It proves especially salient in studies surrounding emotion-filled events. In fact, it provides insights into the impact that research plumbing the depths of difficult circumstances may have on a researcher. As such, hers is an original contribution to the research methodology literature.

Providing examples not seen elsewhere and explaining in detail analysis processes, instead of offering a "black box" approach, strengthen this text. In particular, her explanations of research activities help connect theoretical ideas to on-the-ground research decision making. Performing research that explicitly seeks to open to examination deep emotions, such as Carolyn's did, requires something of a fieldwork tightrope act. Carolyn's research about parents' experiences of the Columbine attack provides a wealth of insightful examples, and she offers others, including studies of Latina/o immigrant parents' talking about their relationships with schools, both in the United States and in their Mexican homeland (demonstrating research strategies inspired, in part, by Mears' dissertation). Providing in-depth examples ultimately improves the likelihood that other researchers will be able to gain deep insights without invading the privacy of participants, or otherwise putting them at risk, as well as learn about data presentation modes that better capture complex circumstances.

In addition, this text provides a succinct chapter about ethical responsibilities of researchers that aligns with federal guidelines used by institutional review boards (IRBs). I anticipate that this will help novice researchers move more quickly through IRB review processes and to accept these as more than simply procedural hoops. Through my service on one of my university's review boards (for behavioral research), I have ample evidence of the need for such a chapter. I found it remarkable that her chapter seemed so relevant to my campus, which is a different sort of place from the one where she earned her doctorate.

Finally, I am delighted to see an appendix that brings technology into the doing of research using qualitative data. Interview research texts and articles rarely encompass the messy nitty-gritty of selecting recording devices, deploying computer software to support transcription activities, and using Web-based resources. Her advice will provide a starting place to novice researchers, as well as to old-timers needing to improve these skills. In addition, she provides a comprehensive annotated bibliography of ethical research practices, interview research, and qualitative inquiry.

Thus, readers will find in what follows a text written in a straightforward fashion intended to guide interview research. As Mears suggests,

> Learning from the life experience of another is a privilege, and while connecting across boundaries of understanding can be a soulful endeavor that privileges us all, it is not easy work. An ethical purpose to drive the research, a tolerance for ambiguity and uncertainty, a willingness to explore unfamiliar and perhaps shadowed terrain, and a commitment to the highest standards of research practice will be required, and that's just to get you started. (p. 145)

When taken as a whole, *Interviewing for Education and Social Science Research: The Gateway Approach* opens research methodological spaces and expands what researchers can know about their participants, for valuing what have historically been seen as issues that were off-limits to research and for conducting interview research that pays attention to the complex, sometimes illogical or self-contradictory, inner workings of human behavior.

Karen L. Tonso, Ph.D.
College of Education
Wayne State University, Detroit, MI
Chair, AERA Qualitative Methods SIG Dissertation
Award Committee
October 29, 2008

Acknowledgments and a Note of Confession

I sit at my keyboard this morning, trying to find the words that might in some small way communicate the enormous gratitude I feel for those who have helped in the preparation of this volume. I have to admit, when I pick up a book, anxious to begin reading, I always skip over the acknowledgments and dig right in to the substance of the text. I don't want to be slowed down with the words of appreciation to people I don't know.

Now that it comes time to write my own acknowledgments, I see the matter in a totally different light. I feel that I need to offer my apologies to all of those other authors whose expressions of gratitude I have slighted.

In many ways, the creation of a book is a solitary journey that is supported by multitudes. No one—well, at least not I—can bring a book to life without the help of many others. I feel incredibly blessed in this, since those who have helped me on this journey are amazing souls—family, friends, colleagues, even strangers who have seen value in the effort and given the gifts of their time, energy, and insight. But before I try to name those who have honored me with their support, since you may not read through the remaining paragraphs, I simply want to acknowledge that this work represents the contributions of many, and to them I say, *thank you.*

Words are such ineffective devices, presuming to communicate what only the heart can say. Recognizing that my words of appreciation will fail to express the depth of my meaning, I humbly acknowledge the following:

My husband Connally, who understands my need to stay focused at the expense of vacations and evenings out and who encourages me when I stumble, celebrates with me when I succeed, and is always there to show me that I am loved.

My sons Austin and Brian, and their families, Christen, Jenna, Ember, and Talon, who are the inspiration for the work, for it is their generation that can benefit from research dedicated to the trite yet driving goal of possibly helping to make the world a better place.

My mother and father, who passed away before this book was begun, but who taught me the importance of education and the ethics of responsible living.

My friends Rose Clary, Sharon Hoery, Paula Kinkaid, and Lynn Melena, who have commiserated with me in times of struggle and walked with me on unmarked and sometimes difficult roads toward discovery and understanding.

My friend Geri DiPalma, who challenged me to write this book and then gently nagged at me until I did.

My friends and colleagues Mary Taylor, Hilary Greenebaum, Marian Bussey, Susan Korach, Bruce Uhrmacher, and Karen Tonso, who graciously joined in the core of readers to review early versions of this manuscript and offered wonderful suggestions for revision and reorganization.

One who was a stranger but is becoming a friend, Corrine Glesne, who responded to my e-mail and instead of deleting it as *spam* generously volunteered to review and comment on the manuscript.

My dissertation committee at the University of Denver—Elinor Katz, Cynthia McRae, Bruce Uhrmacher, and Mary Taylor—who accepted and encouraged my nontraditional approach to researching a difficult subject.

Members of the American Educational Research Association's Qualitative Special Interest Group, who saw merit in my approach to researching the Columbine tragedy.

Julia Cohen, Samantha Hasey, Rachel Tekula, Vivek Muralidharan, and others at Palgrave Macmillan and Integra, who gave this book a chance and shepherded it through to completion.

And, at the base of it all, the Columbine families and others who have placed their trust in me and allowed me to record their stories for my research.

USER'S GUIDE

Interviewing for Education and Social Science Research: The Gateway Approach is a guide to in-depth interviewing for the purpose of creating a gateway to deeper understanding of the complexities of human experience. Interviewing is a component of many qualitative methodologies, either as the primary mode of data collection or as a method to clarify or triangulate data gained through other means. While there are differences in its practice in different traditions, certain fundamentals to interviewing cross methodological boundaries. As a result, this text will be of use not only to researchers who want to conduct a study modeled on the approach I call "gateway," but also to researchers and writers in other traditions as well.

Before going further, I want to offer a brief note on style and voice. I am writing this book with you, the reader, in mind. I'd like to think that we are having a quiet conversation about a practical approach to using interviews in your research. My style and tone may be more informal than you are used to, and that may be disconcerting at first. I hope you will bear with me on this. Some of my earlier readers have told me that they appreciated the accessible nature of the book. Others have found it too informal for a text on scholarly research practice. But, since I am writing about using stories to explore and reveal human experience, a distanced, academic voice just seems plain wrong. As a colleague in your world of qualitative inquiry, one who shares an interest in learning about the human response to experience, I want to pass on to you some lessons I have learned through my own research journey and offer some strategies for you to consider as you proceed with yours.

The in-depth interview approach described in this book is the unanticipated by-product of my investigation into the impact of the shootings at Columbine High School. While I developed the approach to meet the particular demands of my own research, the fact that I am often asked to teach others how to design such a study has shown me that it has utility for investigating a wide variety of situations and events. As an educator and parent of a student who was exposed to the April 1999 attack, I began the research because I wanted to collect the stories of the experience of Columbine parents so that lessons could be learned about recovery from a community-wide tragedy. After I completed the study, I proceeded to write and present on what I had learned. Audiences were appreciative of the findings, but the most common reactions were, "How did you do it?" and, "Is it something you can teach me to do?" It seemed that while the study itself was interesting, there was an equal, if not greater, interest among researchers in every audience in learning how to use the approach for their own investigations. I had not started out to develop a new approach, but as I problem-solved my way through the research process, that is what happened.

In addition to the dissertation, two of my recent publications have contributed to the structure and content for this book. One, published in the *International Journal for Qualitative Studies in Education* (2008), describes the methodological foundations of the approach. The second, published in *The Oral History Review* (2008), explains the potential for positive outcomes for participants in this type of study. Building on these frames, this text traces the development of a model for in-depth interviewing and suggests specific strategies for bringing data to life while ensuring that key research purposes are achieved. It addresses the particular challenges that led to this innovation and examines dispositions that are helpful for anyone considering interview research. Further, it delineates the process, suggests resources, provides a mechanism for overcoming subjectivity, and offers suggestions for researching issues of a particularly sensitive or troubling nature.

I have designed this book for use by a variety of audiences, including graduate students, education researchers, social scientists, oral historians, independent investigators—in fact anyone who seeks to document and learn from the life experience of others. It can be used in many different settings, ranging from introductory research courses, to community history projects, to advanced dissertation design seminars. My goals for the book are as follows:

1. To explore the practice of interview research;
2. To introduce helpful dispositions and practices for interview research;
3. To provide guidelines and standards for designing high-quality, interview investigations;

4. To prepare readers to conduct research using the gateway approach; and

5. To suggest tools and resources that can help in the design and completion of a study using the gateway approach.

The text's organization models the development of a research project. If you are an experienced researcher, you might be tempted to move ahead to the "how-to-do-it" sections. However, I encourage you to start with the introductory chapters, which are designed to provide you with a context for what is to follow:

- Chapter 1 provides a brief look at the origins of the approach.
- Chapter 2 considers core questions related to the choice of any interview methodology, namely the determination of its purpose, driving research questions, challenges for the researcher, and standards for assessing quality.
- Chapter 3 addresses the questions of ethical practice and protection of participants.
- Chapter 4 examines the aspects of oral history interviewing, educational criticism, poetic display, and member check that are interwoven in a gateway investigation.
- Chapter 5 considers the essential process of preparing for gateway research, which includes building a conceptual framework, developing an interview guide, selecting narrators, scheduling sessions, and getting ready for the interview.
- Chapter 6 explains the process of in-depth interviewing to discover the hidden, internalized world of human response and the meaning that people take from an event or circumstance.
- Chapter 7 considers the process of interpretive data display, narrator check to confirm understanding, analysis, and reporting.
- Chapter 8 suggests possible applications and studies well suited to the gateway approach with special consideration for investigations into topics requiring heightened sensitivity.
- In the appendixes, you will find suggestions for technology-based resources and digital tools, samples of forms for data collection and management, and examples of narratives from gateway studies.

I wish you well as you begin your research journey. We can learn much from one another, and it is my hope that this text will prove useful in your search for deeper understanding that can be shared with others.

CHAPTER 1

ORIGINS OF THE APPROACH

CHAPTER TOPICS:

- the Columbine study
- research as a gateway

A LITTLE BACKGROUND

The story behind this book began on a glorious spring day, April 20, 1999, in Jefferson County, Colorado. I had just completed a meeting with educators at a high school in a nearby school district. At the time, I was helping to coordinate a grant-funded project designed to bring inquiry-based teaching to K - 12 classrooms in the area.

As the meeting adjourned, I received a phone call from a friend who very gently said, "Carolyn, there's been a shooting at Columbine." She was concerned for the safety of my son, a sophomore there. Confident that there was little chance that my son would be in danger, I calmly drove to the school—after all, I thought, Columbine was a safe place. Traffic was blocked at an intersection just east of the building, and I was directed toward a staging area at an elementary school nearby, where parents and families were being asked to wait. Three hours later, after a SWAT team freed my son and his friends from the school, our family was reunited. The next day, we learned that two Columbine seniors, Eric Harris and Dylan Klebold, had murdered 12 students and 1 teacher, wounded 23, and then taken their own lives in the school library.

Thus began a nightmare, one that affected my son directly, my family personally, and my community collectively. The struggle through the day-to-day impact of this tragedy was marked by great uncertainty, for each day carried the potential for more turmoil. Indeed, the trauma and loss of April 20th were only the first of many painful challenges. It became difficult to believe that the world would ever be stable, safe, and predictable again.

For me, this experience, this struggle to tomorrow, was achieved with the help of family, friends, community, and even total strangers. For some in the community, relationships deepened as friends supported each other, sharing collective and personal sorrows and triumphs. But for others, relationships became strained and difficult, since all responded to the event differently, constructing meaning in their own individual way. The fact that there was not a single event called "Columbine," but a series of situations and reactions that followed, meant that there were many different experiences, prompting many different responses.

When I listened to speakers at national conferences present their conclusions about the shootings or when I read in professional journals about research on the effects of school violence, I realized that people were trying to understand the situation but with limited knowledge of its complexity. It was clear that the stories that were being shared and the consequences that were being lived out within the Columbine community were unknown elsewhere. In fact, when I talked with other parents about what was being suggested by the national experts or advised by friends in other communities, we frequently concluded, "They just don't get it."

I decided that one way others might "get it" would be for someone inside the community to take on the task of communicating across the boundary of experience. It seemed to me that by sharing the stories of the effects of rampage school violence and the strategies that facilitated or inhibited recovery, I could help educators in other schools and residents of other communities become better prepared. By hearing and sharing the experience from the inside perspective, others could know what to expect if such a tragedy were to shatter the peace of their community.

I wanted to collect stories of the Columbine experience, not in the way that a journalist might report the event, but for the purpose of documenting the elements of the tragedy and then analyzing what was being expressed so that lessons could be learned. I was not interested in constructing timelines or analyzing causes and identifying preventative measures. My goal was to reveal the changes that the tragedy brought into the day-to-day-life of families. I wanted stories to serve as a source of information about what it meant to have this experience so conclusions could be drawn that could help make a difference for others who might have their own world rocked by devastating loss. I realized that to do this, I needed to do more than collect stories. I needed to conduct research.

To prepare for this effort, I enrolled in a doctoral program at the University of Denver College of Education so that I could complete dissertation research on the aftermath of the shootings (Mears, 2005). Other Columbine parents saw the benefit of attempting to learn from the tragedy and offered their support.

CONDUCTING THE COLUMBINE STUDY

Investigating an event that a member of my family had personally experienced brought a particular set of challenges, for insider research understandably raises questions of researcher bias and personal agendas. However, I knew that as a long-time resident of the Columbine community, I had access to information that might be denied to outsiders. It was this inside information that I believed could contribute to the knowledge base about rampage violence, its impacts on individuals and communities, and the measures that might be employed to help in its aftermath.

After a traumatic event, those affected tend to surround themselves with a protective boundary, what psychologists term a *trauma membrane*, in order to avoid further violation (Lindy, 1985; Lindy, Grace, & Green, 1981). In the aftermath of a trauma, such separation is created by the shared experience of intense disruption, with those who are impacted on the inside feeling distanced by their own experience from those outside, who have not lived the event and hence, it is believed, cannot possibly understand a reality that mere words fail to express (Felman & Laub, 1992). This phenomenon helps explain some of the obstacles that researchers face when trying to learn from a catastrophic event. Thus, a major challenge to researching life-changing events is to investigate the experience without being intrusive and without negatively affecting participants who are already feeling victimized and misunderstood.

Recognition of a "membrane" of separation is particularly apt when studying communities affected by a disaster or traumatic event, yet the boundary of experience and understanding is a useful metaphor to keep in mind when investigating other situations as well, whether of traumatic origin or more mundane. Even without the emotional and psychological consequences of exposure to trauma, a separation exists between those who have lived an experience and those who have not. One who investigates the effects of a particular social phenomenon or educational policy, for example, will bring a different lens to the study if prior experience is involved, but this does not mean that only insiders can research the phenomenon or situation at hand. What it does mean is that the researcher needs to have sufficient background knowledge and affinity for the topic to engender the trust of the study participants, to build the genuine rapport required for successful interviewing, and to understand the "language" of the experience in order to discern the significance of what is being said.

Successful interviewing requires attention to another's point of view. Establishing the connection and rapport that make this possible requires "becoming informed about your setting's social and political structure so that you can shape your conduct with sure-footedness that such knowledge affords . . . it is the knowledge that helps you fit in" (Glesne, 1999, p. 101). It is also the knowledge that helps you accurately interpret and clearly understand what you hear.

The value of prior knowledge or experience must be weighed against the potential for bias. Finding that balance and achieving a "virtuous subjectivity" (Peshkin, 1988) requires diligence, self-reflection, and commitment to disclosure. It is not possible to guarantee absolute objectivity in research, for indeed, whatever the topic, it would seem that the researcher must feel some subjective affinity to that area of inquiry; otherwise, the matter would hold no appeal. What is vital, though, is to acknowledge the presence of a subjective lens and assess its impact throughout the research process.

For the Columbine study, I committed to maximizing the strength of an insider perspective while minimizing the negative impacts of subjectivity as I answered the research question, *What is the experience of parents whose children have been exposed to a school shooting?* I did not want to take on the risk of interviewing students who had faced the gunfire, but I felt that I could learn from the parents' perspective what had helped their family move toward recovery.

From the start, I acknowledged that the topic was not one that I could approach with detached objectivity. But my purpose was not to compose a personal essay, a telling of my own story. Instead, it was to explore the experiences of others so that lessons might be learned about planning for and responding to a traumatic event in a school or community. The study required vigilance throughout, in monitoring my own responses, designing safeguards to protect the study participants, conducting the interviews, presenting the data, drawing conclusions, and reporting results.

Even with a clear purpose, audience, and research question, this journey posed dilemmas that forced me to make difficult decisions about how to proceed. Among the first of the challenges was to identify a method. I knew that a qualitative approach would be most appropriate, but I soon realized that I could design the study in a dozen different ways. My search for stories aligned with oral history, but I did not want to write an oral history of the Columbine tragedy. I could have researched the culture of a community after a tragedy (ethnography), attempted to come to terms with the phenomenon of school shootings (phenomenology), or investigated this as a case study of a rampage shooting in a large suburban high school. I considered these and other approaches, but none quite captured the essence of what I wanted to do. I wanted to connect those outside of the event directly to those inside, communicating the experience in a way that would evoke a depth of understanding, with me, the researcher, as invisible in the process as possible.

No standard methodology seemed to quite satisfy my goals for the research and the particular concerns related to conducting research within a traumatized community. I was fortunate to complete my doctoral studies at the University of Denver with advisors who understood as I faced challenges that could not be met to my satisfaction through traditional means. Not all students are so lucky.

In order to safely capitalize on an inside perspective while avoiding bias that might skew the study, I selected elements of open-ended oral history for interviewing and Elliot Eisner's (1998) arts-based approach to educational criticism for data analysis. (You will learn more about these approaches in later chapters.) After I had completed a series of interviews with each narrator, I began the task of sorting and learning from mounds of data. At this point, I realized that I simply could not do justice to the participants if I merely summarized or paraphrased their words for consideration. I did not want to just tell what happened to them and their family. I wanted to reach across the trauma membrane and evoke an understanding of what it felt like to actually be living the immediate and ongoing consequences of this experience. This meant that I needed a way to communicate the emotions, the fears, the confusion, the sense of loss, and the many other responses that are brought on by exposure to trauma. I felt that knowledge of the complexity of the situation would be prerequisite to designing productive strategies to assist in the recovery.

Because of the diversity of responses to the tragedy, it was important that I represent the parents' experience in their own words and context, and not just interpret and report it through my lens. To promote authentic understanding of these intense personal experiences, I needed to connect the reader directly to what the parents had to say. I wanted to humanize the experience so that others could understand the complex circumstances that develop in the aftermath of a life-changing event. I could have just analyzed the transcripts, listed specific effects of the tragedy, identified factors that promoted recovery, and drawn conclusions. However, for the deeper human consequences of an event to be revealed, the voices of the participants needed to be heard. I was at a loss for how to do this, but fortunately stumbled on a technique developed by Laurel Richardson (1992) and adapted by Corinne Glesne (1997) that provided a model for excerpting transcripts into a narrative form that comes to life with poetic vibrancy. Adapting this technique allowed me to keep the narrator at the forefront of the research, establishing a context for the conclusions and bringing a deeper understanding of their significance.

Having settled on a way to work with the data, I then needed to make sure that I did not change the narrators' meaning in any way and did not overlook what they considered most significant. I knew that member check would be important to confirm the accuracy of the transcripts, but as a key element of the design, I asked narrators not only to review the transcripts of their interviews but also to read and evaluate my interpretation of their individual narratives. This essential check to clarify and confirm the record provided an opportunity for the participants to examine their own thinking and to recognize and reflect on their own life experiences. This was one of the most informative steps in the process, for it allowed me to encourage the narrators to go beyond what had been included in the original interviews, in essence reflecting on their reflections.

Throughout this investigation, the research purpose drove the method instead of the other way around. It might have been easier to simply select an established methodology and fit my work into that structure, but that would not have fulfilled my purpose for undertaking this effort. When I had finally problem-solved my way to completion, I began to present findings from the research in conferences across the nation. The surprise for me was that in addition to taking note of the recommendations from the study, audience members always asked me to tell them about the approach, since they were interested in using the process for their own research. The remaining chapters of this book give that information, but it seems appropriate at this point to provide you with a brief example. Here, I'd like to demonstrate one aspect of the approach, the data display model, which I adapted from a strategy of excerpting phrases from interviews to create narratives that take on a "poetic" form (Glesne, 1997).

The following excerpt demonstrates the power of presenting interview data in story form, for it provides a context that brings to life the underlying reasons for a simple recommendation for promoting recovery. Presenting the first-hand experiences in this manner evokes immediacy and resonant understanding. Stating a distanced recommendation, perhaps saying that parents should go back into the building with their children after a school shooting, fails to communicate the intensity and context behind that conclusion as is apparent in the following excerpt.

Excerpt from Rebecca's Story

Articulate and reflective, Rebecca thoughtfully spoke of her experience as a mother of two sons enrolled in Columbine at the time of the shootings. At her request, her first interview was held in my home. Her second and third interviews were held in her home. In both settings, she devoted full attention to the questions, pausing occasionally to organize her thoughts before answering softly and insightfully. Rebecca is an elegant and precise speaker; her replies were rich with metaphor and descriptive detail, reflecting the breadth of her education and extensive reading. Describing the terrible pain that strikes a community that experiences the type of loss incurred in a school shooting, Rebecca offered solace, reminders of the importance of hope, and the transformative nature of the trauma. Following are excerpts, in her own words, from Rebecca's story:

Returning to Columbine . . .

They got the school cleaned up,
And one night they opened it
For kids to take their parents in.
We went and
We saw where our child had been.
That was very important for us.
You have all of these questions as a parent,
As to what your child experienced.
So you want to understand it.
You have an overwhelming need to know.
You can't ever experience what they experienced,
But you had a need to understand it.
Somehow seeing the place helps you visualize
What it must have been like.

I wanted to go back in,
To see the place,
To look out at what he looked at
To sit where he sat,
To stand in his shoes for the three-hour period that he was in that room
So I could imagine the fear.
That is what my child experienced and
I wanted to understand.

So that's what we did.
We went into the room,
Then we could understand,
To put the pieces of the puzzle together.
Someone had placed a rose in front of the door where he had been.
It just represented the pain.
It was a tribute to the pain.

What was important is that we went in with him,
And he pointed out,
This is where I was,
This is what I saw out that window,
The sharpshooters were over there,
But we didn't want to stand up.
We didn't know who was going to see us.
Maybe the shooters.
And this is what we could hear outside of the door.

It was so important that he share it with us as his parents.
Then we went to the other rooms to see
What other kids had experienced.
And of course we walked by the library.
It was sealed off, but it felt like a tomb.
It was like sacred ground.

(Mears, 2005, p. 133)

This type of narrative display provides a medium for confirming that the story reflects the narrator's meaning and, equally important, it does so in a way that communicates the emotional and psychological responses, not just the physical events or actions. By preserving the narrator's voice and confirming the deep levels of meaning, the presentation becomes very real and immediate, "giving a face" to the research (DiPalma, 2007). It evokes a deep understanding of the complexities that lie within the experience itself, the basic human responses, and the individual interpretation of its significance.

Before the participant's voice can be preserved, it first has to be heard and comprehended. In-depth interviewing from an informed perspective provides an opportunity to learn lessons from experience that might otherwise be lost. In conducting the Columbine research, I was keenly aware of the ethical imperative to protect the study participants from harm and exploitation. At the same time, I felt the profound responsibility to preserve and accurately portray the participants' perspective and understandings in the research.

By agreeing to participate in any study, narrators are in fact granting entry into their world of experience. As a guest in that world, a researcher needs to be faithful and accurate in representing what is seen and heard and learned in that setting. It was my commitment to this principle that shaped the study and the development of a distinctive approach to qualitative inquiry.

RESEARCH AS A GATEWAY

In 2006, my dissertation *Experiences of Columbine Parents: Finding a Way to Tomorrow* was recognized for its methodological significance by the American Educational Research Association (AERA) as the Outstanding Qualitative Dissertation for the previous year. The review committee strongly encouraged me to publish and share my approach. The study was commended for drawing on "distinctive epistemological traditions to develop both a methodological approach that was not only consistent with but sensitive to researching and analyzing the experiences . . . of those [in the study]" (Kuzmic, 2006). A member of the review panel wrote to tell me that showing my research to her doctoral students "opened doors [for students] that they have been trying to open for a while." She encouraged me to "keep working at helping us understand how to do this, so that we too can help others hear what they are unaware of or lack access to."

I decided to use the term *gateway* in describing the approach, since it provides a means of connection, a way toward deeper understanding of a metaphorical "community" of experience. With direct access to the complex internal responses and the significance that study participants take from their life events, those *outside* of that experience have the potential to achieve a deeper understanding of what it feels like to be *inside*. Similarly, the gateway approach provides a means for those *inside* to cross the boundaries and communicate with those *outside* who want to learn from the situation. In addition, participants have reported that the process offers them an opportunity for personal reflection, and in this regard it may serve as a pathway to increased self-understanding and empowerment. As one of the participants in the Columbine study noted,

> You should learn from [an experience]
> I've shared some things with you
> Whenever you share,
> You have a better understanding of things.

The gateway model grew out of my research into the aftermath of the Columbine tragedy. Although this approach is particularly useful for investigating the physical and emotional consequences of experiencing life-changing events, it has been recognized by other researchers as having the potential to deepen understanding of more routine situations as well. Among other applications, it has been used in an investigation of innovation and problem solving, a review of challenges faced by mid-career doctoral students, a proposed analysis of kindergarten literacy pedagogy, and a study of public policy and families of children with special needs. Its techniques for data display and confirmation have also been used in a study of involvement of parents of immigrant children in urban schools.

In my dissertation research, I faced the struggles normally encountered in academic work, plus a few extra that came along with researching within what was considered a high-risk setting. In the process, I found a way to achieve the goals that I had framed from the start. I had hoped to collect and share stories that reflected the wholeness of the experience, bringing the narrators into being as complex, living individuals, in a holistic context—not just presented through disembodied facts, timelines, or summaries from a distanced perspective. I wanted my research to have the depth that makes the findings understandable on a human dimension. The solutions that I found have been employed by others with much success. Perhaps they may be of use to you as well.

WHAT'S NEXT?

Since interviewing is the organizing element of the gateway model, I have decided to begin the discussion with an analysis of the nature and ethical practice of in-depth interviewing in general. These concerns are common to interviewing methodologies across the qualitative spectrum. Following the discussion of these general guidelines is a consideration of the foundations of the gateway approach, the specific processes that it involves, examples of its practice, areas of research where it might be applied, and resources for interview research.

I hope that you see value in creating a gateway into a world of experience from another's point of view. You may choose other avenues for your research, perhaps building on this starting point to find your own point of entry. It is up to you to find your own way through the challenge as you complete your own research journey. I'd love to hear your story.

CHAPTER 2

THE NATURE OF INTERVIEW RESEARCH

CHAPTER TOPICS:

- storytelling
- the practice of in-depth interviewing
- research purpose and questions
- helpful dispositions and interests
- challenges, criteria, and standards

This chapter considers the common elements of in-depth interviewing for a research purpose. These general traits, which are characteristic of most interview investigations, are a starting point for interviewing for gateway research into the many layers of significance and meaning that arise in response to an event or experience.

Interviews that solicit stories of personal experience offer a powerful point of entry into a world from another's perspective. When I began the Columbine investigation, I knew that I wanted to collect the stories of parents whose children had been exposed to the attack, but I wondered if story-collecting could be rigorous enough to be considered research. Certainly, the collection of life stories has value to historians, genealogists, and others dedicated to documenting lived events. But in education and social science research, sophisticated measurement tools and technologically based procedures can expedite data collection and analysis in ways that appear to generate more dependable results. In an era in which researchers are pressed by demands for scientifically based research and a "predisposition toward the glorification of the work of scientists and the technologies it produces" (Barone, 2007, p. 455), the time-honored traditions of storytelling that once brought together ancient peoples around stone fire-pits would seem to be of questionable value. Yet, it is perhaps the nature of those ancient echoes that gives power to the story and allows the expression of significance and meaning to be heard on a deeply resonant level, a level at which experience can be witnessed and outcomes understood.

STORYTELLING AND RESEARCH

Stories can be so much more than simple chronologues of *whats* and *whens* and *what ifs*. In the simplest of terms, "we know the world through the stories that are told about it" (Denzin & Lincoln, 2005, p. 641). To hear someone's story is to enter a world of thought, action, emotion, and circumstance through another's perspective. Sharing one's story requires reflection and discernment of what is meaningful, significant, and safe to tell. "When people tell stories, they select details of their experience from their stream of consciousness. . . . It is this process of selecting constitutive details of experience, reflecting on them, giving them order, and thereby making sense of them that makes telling stories a meaning-making experience" (Seidman, 2006, p. 7).

In-depth interviewing offers researchers a method for accessing stories to broaden understanding while honoring the authority that individuals have over their own life memories. Storytelling and story-hearing offer a meeting ground for deepened connection, clearer understanding, and mutual learning. From that learning, comes the potential to bring change when change is necessary, and to support preservation when it is not. To communicate a world of experience so that appraisal of impacts and implications for action are possible requires that the story of experience be accurately heard and documented and then shared in ways that contribute to deeper understanding by others.

As narratives of experience, stories present the "possibilities and limits of what people may do in similar circumstances, even when we cannot predict what they will do. By indicating what might happen, stories enable us to prepare for a range of eventualities" (Stiles, 1993, p. 601). The stories that rise in the memory reveal what holds particular significance for the individual at that particular time. They ground the research, providing the context and conditions that underlie behaviors and beliefs to reveal what the narrator perceives as noteworthy enough to share. In an interview, people have a chance to reflect on their experience and through that process consider their memories with intention. Narrators make choices about what to share with an interviewer, and these choices are connected to perceived significance, with details provided or omitted to support the telling of the memory.

Interviews collect more than details about an event. "Neither contemporary nor historical evidence is a direct reflection of physical facts or behavior. Facts and events are reported in a way that gives them social meaning" (Thompson, 2000, p. 128). Interviews thus offer access to that place where interpreted human experience and response intersect with an educational, social, cultural, spiritual, or political dynamic, providing the means by which privately held contents of memory can be communicated to a listening researcher. Further, as Portelli (1991) has observed,

> What is really important is that memory is not a passive depository of facts, but an active process of creation of meanings. Thus, the specific utility of oral sources . . . lies, not so much in their ability to preserve the past, as in the very changes wrought by memory. These changes reveal the narrators' effort to make sense of the past and to give a form to their lives, and set the interview and the narrative in their historical context. (p. 52)

Psychologists recognize that "memory is not a literal reproduction of the past but instead depends on constructive processes that are sometimes prone to errors, distortions, and illusions" (Schacter, Norman, & Koustall, 1998, p. 289). Schacter (2001) used the term *sins of memory* to describe the traits of omission (transience, absent-mindedness, blocking) and commission (misattribution, suggestibility, bias, persistence) that commonly occur in this process. These factors play into the retention and attribution of meaning; thus the details that are recalled when processed through the lens of human memory reveal a great deal about the meaning-making process itself.

Once experiences are reflected on and told in an interview, the human dimension of an event or experience begins to take shape. Abstract findings regarding the implementation of a political or social policy, for example, become powerfully real when an individual's account of that policy's effect on his or her life become clear. The details of the policy implementation need to be verified elsewhere, but the individual's perceptions about it, its benefits, the challenges, and the significance it has—to that very real individual—can only be accessed through in-depth interviewing.

My own interest in learning from the way individuals and groups experience their life events leads me to ask the kinds of research questions that can best be answered through this process. The quest to explore and learn from experience and then share understanding begs for stories and sounds and expressions, not numbers and ratiocination. The art of accessing and learning from the reflections and memories of others is one that builds with the slow, steady collection of stories through the sensitive yet complex act of the in-depth interview. Interviews offer opportunities to cross boundaries of understanding and to learn from the behaviors and life events of others, uncovering insights from the impacts of a situation, or a program, or a policy as revealed in human terms and then communicated in ways that can be used by the people who create the situations, or design the programs, or write the policies. My reason for conducting research, quite simply, is to bring people closer to an understanding of the experiences of others so that situations and programs and policies can be better understood and, if necessary, improved.

When I first became involved in academic research, I started from this point of self-awareness. I knew what I wanted to do and then looked around for tools and strategies that would help me achieve my goals. For me, part of the answer lay in in-depth interviewing. Before you begin your own research, you will need to consider your personal goals and interests that will affect how you conduct the study. The remainder of this chapter can help prepare you for this essential step.

WHY, WHAT, AND BY WHOM?

Your research purpose and questions determine which approach will be the most productive and appropriate for your investigation. A careful assessment of your goals, specifying the questions you hope to answer and how those answers could be put to use, can help you avert a methodological mismatch. No single approach can be made to answer every question and serve every need. Ultimately, the first matter of business is not to select a method, but to define a purpose and then consider how best to achieve it. Designing research also requires an element of self-reflection. Thus, research begins with a consideration of three factors that influence the choice of a productive methodology for the investigation:

- the research purpose (how will the findings of the research be used),
- the research questions (what questions does the researcher hope to answer), and
- the personal disposition of the researcher (what researcher traits support achievement of the methodological demands).

Before you begin your research, it is important to address these fundamental considerations. The first two, the identification of purpose and questions, you may have already considered in research design classes, but the third matter is often overlooked. A reflection on your own interests, likes, dislikes, talents, and limitations can help you design a study that achieves your research goals with a minimal amount of personal anxiety and struggle.

In any investigation, research and researcher must interact. I realize that this sounds like a simplistic observation, but the truth of this matter is often not discovered until much too late in the process. A method that cannot answer the research question is no greater a mismatch than a researcher whose disposition cannot acclimate to a mode of thinking that the method requires.

Research purpose

As you begin your research, clarify your purpose and be able to state with conviction, why you are doing what you are doing. Now, if you're a graduate student reading this, your first response might be, "because I have to." Certainly, if you're working your way toward your doctorate, that little matter of the dissertation could be a key motivator. But that's not the *why* I am speaking of. The purpose of your research is much more than that.

Even if it's simply a matter of checking off one more degree requirement, you need to design your work in a way that addresses an audience and provides them with the findings you think they need to know. Defining the purpose of your research clarifies how you envision it being used. What meaning do you want it to have? Who, beyond your advisor, will be interested in reading it? Who can work or live better if they learn from the results of your study? Are you looking to inform policy makers? Practitioners? Historians? The public at large? Do you want to influence the world of educational planning? Could your investigation be useful to curriculum developers, or social workers, or teachers? Could your findings have utilitarian value? Could they broaden a philosophical understanding? Could they empower advocacy?

Conceptualizing and verbalizing your ideal purpose creates a touchstone that can help you make the challenging decisions you will face as your research proceeds. An ethical purpose, clearly framed as the foundation for your work, helps guide your decision making as you move along. When you can state a purpose for your research, how it could be used, and why you are committed to the effort, it becomes easier to determine the questions you need to ask in order to inform those who could benefit from knowing the answers.

Research questions

Since interviewing provides access to experience and responses from another's perspective, it is particularly appropriate as a means to investigate how people are affected by events or situations; to discover their thoughts, their feelings, and the meaning that they take from their experiences; and to connect these meanings to their world and to future actions. In-depth interviewing allows the exploration of how ideas and emotions about events and people change through time and experience. Interview-based research involves the interpretation of the story that is being told, with all of its complexity, ambiguity, and even inaccuracy. The interviewer and the narrator share this interpretive process, since the story being told belongs to the narrator, while the telling of it to another brings it into being (Frisch, 1990).

Research questions related to the *whats* and *hows* of human experience seem particularly appropriate to investigate through in-depth interviewing. For example, a research investigation to analyze the effectiveness of an accelerated program for gifted students in terms of academic achievement might best employ a quantitative, statistical approach. However, if the researcher wants to learn *what* the actual experience of students in the program was, perhaps attempting to discover *how* the program impacted their social development or brought changes in their behaviors and attitudes about learning, then in-depth interviews would be a preferred approach.

As a type of interview that records a person's life history or an experience related to a particular event or situation, oral history offers techniques that can be employed in more general in-depth interviewing to great advantage. The open-ended nature of oral history questions invites narrators to focus on what they consider to be important and gives them the power to control the content and the scope of the interview (Yow, 1994). As a result, it provides a powerful strategy for uncovering *what* happened and *how* it affected the narrator.

Perhaps a simple example at this point can demonstrate the connection of research purpose to research question and research approach. In Table 2.1, notice how different research questions can contribute to achieving the same purpose (improving after-school programs), depending on how the terms are defined. In the first example, the research question addresses *improvement* as defined by the number of students participating in the programs. In the second

Table 2.1 Research purpose, question, and outcome/product

Purpose for doing the research: To provide district-level administrators with needed information in order to improve after-school programs

Research question	Potential data collection methods	Possible outcome/ product
1. Which after-school programs are most likely to be supported by middle school students?	Student survey Attendance data Feedback forms	List of options and priorities
2. How does participation in after-school programs affect the student's academic achievement?	Records review Achievement data Observation	Documentation of quantifiable changes in grades and behavior
3. From the student's perspective, what are the benefits or impediments to participation in an after-school program?	In-depth interviews	Disclosure of student perception of advantages as well as barriers to participation

example, the researcher wants to know if programs affect the students' academic achievement. In the third example, the researcher wants to learn about the student's experience, so that programs could be designed to maximize benefits while minimizing inhibiting factors. It is also clear from Table 2.1 that different methods of data collection are required, based upon the research question and the products that result. The researcher needs to decide in advance how terms are defined and how the findings will be used.

Helpful researcher dispositions

When the research purpose and the research question point to a qualitative study using in-depth interviews, the next issue to be addressed has more to do with the researcher than with the methodology. For many novice researchers, the allure of qualitative inquiry may be predicated on the misconception that it offers a "simpler" approach, one that is perhaps less challenging than a quantitative foray into managing numerical data and statistical analyses. As a result, those researchers may settle on a qualitative practice, perhaps collecting data through interviews, since they think their years of experience in the simple art of conversation have prepared them for this work. Then, months down the road, they are dazed by contradictory information, bogged down in a quagmire of seemingly endless verbal data, and lost without a clearly defined footpath to guide them through.

The truth of the matter is that "research—like life—is a contradictory, messy affair" (Plummer, 2005, p. 357). If a researcher thrives on that messiness, celebrates the ambiguities, and enjoys the challenge of finding a path with only a few guideposts, then interview research may be a good fit. If the researcher also revels in learning about the sometimes unpredictable way that people experience the world and the significance that their experiences hold for them, then in-depth interviewing might be a good match. If, however, a researcher has greater interest in analyzing variables to reveal causality or test hypotheses through a well-defined process, then a more structured, analytic model may offer a better chance for success.

Effective interviewing requires that the researcher enjoy interacting with people. The researcher attempts to gain "complex knowledge directly from people with certain attributes or life experiences—knowledge about their experience and the contexts influencing their relations to others, behavioral choices, and attitudes" (Hesse-Biber & Leavy, 2006, p. 4). In-depth interviewing requires that the researcher be interested in making meaning with another, bridging borders of experience that brings "insider" information to the outside so that those who have not lived the experience can better understand. This type of investigation takes place in the natural setting, not a controlled laboratory or experimental environment. An adept interviewer is able to listen intently and hear meaning in another's words, silences, and postures. This heightened attention to detail cannot come, however, at the expense of rapport, for fundamental to a successful interview is the building of trust that connects researcher and narrator so that there is safety in disclosure.

An open-ended interview approach makes it difficult to know exactly where the research will lead. The interview may not proceed as the researcher anticipates; desired information may not be forthcoming; narrators may tell stories of a sensitive or troubling nature, use offensive language, express outrageous perspectives, or even unwittingly tell of illegal activity. A researcher needs to anticipate the unexpected and to plan strategies in advance for dealing with potential conflict and uncomfortable situations. Each study is unique, and a realistic appraisal of possible developments combined with well-conceived strategies for dealing with these contingencies is an integral part of research design.

Successful in-depth interviewing requires that the researcher possesses a disposition to carry out the research as well as the skills and knowledge to serve as the primary mechanism for the collection of data. Interviewing requires that the researcher play an active role, in fact, using "the self as an instrument" (Eisner, 1998, p. 49). The researcher is not only required to assess which informants can best contribute to the study, gain access to those informants, and establish rapport, but also to provide a framework for the interview sessions, ask open-ended questions to stimulate discussion, discern salient points, search for deeper meaning, evaluate and interpret interview data, draw well-founded conclusions, and then communicate those findings to others. During the interviews, the researcher must effectively relate to people while attending to the process. Following the interviews, the researcher must manage and make sense of staggering amounts of data and a variety of analysis strategies without the comfort of explicit procedures and unequivocal rules on which to rely.

Certain abilities serve as assets to help in mastering these requirements. Some of the assets may stem from the personal preferences and dispositions discussed earlier. Others are connected to more discrete, though learnable, skills. If you are considering research based on in-depth interviewing, it is up to you to self-reflect and decide what seems reasonable and achievable. Consider the following list and then assess your own ability to meet these demands. As an interview researcher, you will have to do the following:

- *Enter into the complexity of people's lives.*
  This requires curiosity about human experience, patience, an ability to suspend judgment, and a tolerance for ambiguity and inconsistency.

- *Establish and maintain rapport.*
  This requires interpersonal skills and the ability to achieve productive communication even with people you may not like.

- *Complete a study without definitive steps or authoritative procedures.*
  This requires you to be comfortable in finding your own way without the luxury of explicit directions for the one "right" way to do the research.

- *Ask challenging questions.*
  This takes critical thinking as well as tact in framing questions and the courage to listen to difficult replies.

- *Listen and hear meaning.*
  This requires being able to listen from another's perspective and to understand intended meaning that goes beyond the literal word.

- *Attend to nuances of expression, silences, and nonverbal cues.*
  This means being auditorially and visually attentive throughout the entire interview.

- *Maintain healthy boundaries.*
  This requires a professional demeanor, a healthy sense of self, and personal limits.
- *Engage in self-reflection.*
  This requires that you consider why you are thinking what you are thinking and that you explore your own relationship to the topic and your interviewees.
- *Organize, manage, and analyze an abundance of verbal data.*
  This means that you need organizational and analytic skills as well as the ability to commit time and energy to the task.
- *Find patterns in possibly contradictory data.*
  This means that you must attend to the elusive while not overlooking the obvious.
- *Synthesize information from a variety of sources.*
  This requires accessing, analyzing, and accounting for multiple data sources in discerning effects/consequences on the lives and perceptions of your participants.
- *Write and communicate effectively.*
  These skills are needed for any research endeavor but are especially important for in-depth interview research.

Facing the Challenges

In considering the nature of interview-based research, it is necessary to acknowledge the challenges that it presents. Labor and resource intensive, interview research is characterized by an emerging design and uncharted terrain. It demands time and energy and can require months to complete. Interview-based investigations often raise concerns about scientific rigor and the relative usefulness of findings. These challenges need to be faced squarely so that there will be fewer surprises once you have started your study.

Time, effort, resources

One particularly daunting aspect of interview-based research is a characteristic shared by qualitative research in general. As others have observed, the data are "messy, and usually voluminous. We wind up with huge piles of texts . . . and have to sort our way through them" (Weitzman, 2004, p. 145). The "lack of rules, vast amounts of data to process, the tasks of writing are baffling to some" (Lichtman, 2006, p. 19). Rather than a reliable map with the steps clearly marked and a definitive set of procedures outlined, an interview-researcher embarks on a quest that leads toward a certain destination, but the steps along the way are subject to change with new information gained in any given interview. Even the number of narrators to interview is dependent on a general sense of what will constitute sufficient data. After the very first interview is completed, the time-consuming task of transcription begins. In the Columbine study, while I was delighted in the amazing insights that the parents shared with me, I was overwhelmed at the prospect of transcribing the interviews so I could reflect on and analyze their significance. Some interviewers hire a professional transcriber to transfer the audio recordings to text, but that is an expensive endeavor, especially since a single 90-minute interview can require somewhere in the neighborhood of five or six hours to transcribe. In addition, much can be lost in not transcribing the interviews yourself. If you decide to hire a transcriber, you need to check for accuracy and completeness by listening to the tape while you review the transcript. Whatever strategy you use, plan to invest time, money, and energy in transcription. To solve my own transcription dilemma, I chose to use voice recognition software. (See Appendix A for a discussion of technology tools and other resources.)

After conducting and transcribing multiple interviews for all narrators, the next challenge is deciding how to process and analyze the abundance of data. The Columbine study generated over 600 pages of transcripts. The thought of processing notebooks full of interview transcripts and then reducing that data into manageable form for analysis and presentation in the text pushed me into an almost catatonic state of data overload. It was this challenge that led me to consider alternatives to data reduction and display through the narrative device you will learn of in later chapters. For now, please trust me, data overload is almost a synonym for in-depth interviewing.

Nonlinear design, unpredictability

Even though the process of interviewing for research lacks definitive, absolute rules for how to collect, sort, reduce, interpret, analyze, and present data, assumptions and traditions within the various qualitative methodologies do provide an overall structure and framework in which to work. This text also offers strategies that can help add a measure of predictability to the uncharted terrain. However, the fact remains, interview research rarely conforms to an easily discernible, linear path.

The emerging design of a study may mean that three interviews become four, or that an unanticipated document review must be performed to support follow-up in a subsequent interview session. Factors that cannot be known until the research is begun may complicate the selection of study participants. A single interview may produce contradictory details or insufficient information. The contents of the interview may be interesting but not at all informative or relevant to the question. It may be necessary to recruit additional informants to gain insights from other perspectives to support the research. In fact, data analysis, which in quantitative investigations tends to come after data collection, actually begins in the first interview itself, with the discerning and shifting of cognitive paradigms to accommodate new information as it is received. Thus, the interviewer needs to assess and analyze the status of the data on an ongoing basis, before asking the next question in search of clarification or additional detail.

Measuring up to standards

Research questions that can be addressed by in-depth interviewing connect most directly to the way individuals and groups experience and interpret the circumstances and situations they encounter. Since this type of research is an eminently personal form of data collection and analysis, questions of scientific rigor are often raised.

Validity, reliability, replicability, objectivity, and utility are traditional standards for research, and many investigators have tackled the application of these criteria for assessing the value of qualitative studies. It is generally agreed that in the world of qualitative, interview-based research, these traits mean something a little different from what they might mean to a quantitative investigator. Even the basic understandings of "reliability and validity, as developed for quantitative research, require adaptation for application to qualitative research" (Weitzman, 2004, p. 145). In fact, as Howe and Eisenhart (1990) have observed, for qualitative research

> . . . the question of standards must be viewed wholly within an interpretive perspective, broadly construed. Furthermore, insofar as no standards completely divorced from human judgments, purposes, and values can exist and insofar as there can, accordingly, be no monolithic unity of scientific method . . . standards must be anchored wholly within the process of inquiry. (p. 3)

Individual researchers hold different opinions and perspectives on standards for research. Instead of summarizing them for you here, I offer the following brief points as a way to signal some of the criteria you must consider in designing your study. Your advisors will be able to help you understand how specific research standards apply to your own inquiry.

Validity

Let's begin with the matter of validity, for it is considered a core measure of the worth of an investigation. Validity in qualitative research can be viewed in terms of its credibility, both of its process and its findings. However, an assessment of research in terms of validity "does not lead to a dichotomous outcome (i.e., valid vs. invalid), but represents an issue of level or degree" (Onwuegbuzie & Leech, 2007, p. 239).

What is valid in interview research is the degree to which it illuminates what it claims to inform, what credibly captures and portrays the meaning and significance of representative participants' perspectives on a set of events and experience. This requires an internal consistency that is not necessarily dependent on conformity to an externally verifiable set of facts. In describing interview research in the tradition of oral history, Portelli (1997) observes that it tells us less about events than about their meaning. This does not mean that such research lacks validity, but that it has a different sort of validity. "The validation of oral evidence can be divided into two main areas: the degree to which any individual interview yields reliable information on the historical experience, and the degree to which that individual experience is typical of its time and place" (Lummis, 2006, p. 255). I would add a third area of evidence, namely, that it is either internally consistent or that it addresses and accounts for any inconsistencies that are revealed.

Seeking the narrator's perception of an event or situation, in-depth interviewing explores lived experience and offers "prisms on the past rather than windows" (Heinige, 1982, p. 5). Narrators tell of their own understanding and interpretations of what they remember, and it is not unusual for there to be conflicting accounts and errors in the reconstruction of past events. Instead of basing an assessment of validity on factual accuracy, it is important to consider that inaccuracies may actually reveal a greater truth. For example, Felman and Laub (1992) share the story of a female survivor of Auschwitz, who told of a revolt by prisoners at the concentration camp and related her sense of triumph when she saw the burning of four of the crematorium chimneys. Historians reviewing her testimony argued that her account had little credibility because only one chimney had been damaged in the revolt and ultimately, those responsible were put to death. Her entire testimony thus was cast into doubt. They were demanding factual accuracy while missing the greater truth: For those in Auschwitz, it mattered not if it were one chimney or four. The event itself was perceived as a victory of unimaginable proportions. That even one chimney could be damaged marked a symbolic triumph, and literal accuracy had been replaced with a far deeper significance. The testimony held authority of meaning and witnessed a greater truth than an inventory of chimneys destroyed could ever convey.

Reliability
With factual inaccuracy and symbolic interpretation as distinct possibilities in a narrator's reconstruction of events, questions of reliability are easy to anticipate, for reliability "refers to the trustworthiness of observations or data" (Stiles, 1993, p. 601). This aspect can be assessed by reflecting on the data with an eye to the purpose and focus of the research. If a researcher's purpose, for example, is to produce a factual record of program implementation or historical events, then to be reliable, the data provided in an interview would need to be confirmed, as much as possible, through triangulation with other sources, such as institutional records, public documents, databases, or interviews of other informants. However, when the intention of the research is to uncover the deeper meanings of an experience, what life impacts were felt, what personal significance the event held, and not to establish an exact record of names, dates, and places, then the question of reliability rests with the question of the authority of the narrator, the internal consistency of the story being told, and how well the account agrees with the narratives of others. In fact, even if the account were deliberately given falsely, that rendering in itself is a type of data informing the research. "The importance of oral testimony may lie not in its adherence to fact, but rather in its departure from it, as imagination, symbols, and desire emerge. Therefore, there are no 'false' oral sources" (Portelli, 1991, p. 51). Even an intentional lie is a form of communication, and it is the job of the researcher to figure out what to make of it.

Allowing each narrator to review his or her data (transcript) and the interpreted meaning that the researcher takes from that data contributes to both reliability and validity. When the voices of multiple narrators are represented, it is important to check for the coherence of the data and account for any differences or similarities in the reports. Through a series of in-depth interviews it is possible to check details from across all sessions and then acknowledge and evaluate significant discrepancies in the analysis phase. For investigations aspiring to factual accuracy, Heinige (1982) offers a strategy in which narrators are interviewed and asked the same questions more than once, as a way to check the reliability of their information.

Replicability

Given the variety of narrators that could be chosen, differences in interviewer–narrator interactions, lack of definitive rules for data analysis, and a lack of "operationally defined truth tests to apply to qualitative research and evaluation" (Eisner, 1998, p. 53), establishing replicability of an interview-based study requires meticulous attention to details and process. Replicability, which is not a quest to reproduce the research with perfect exactitude, is measured relative to the transparency of the procedures, consistency of the work, reasonableness of results based on the interview data, and disclosure of the researcher's connection and subjective knowing that may have influenced the conduct of the study.

Eisner suggests that the criteria for replicable research include *consensus* of other investigators "that the findings and/or interpretations . . . are consistent with their own experience or with the evidence presented" (Eisner, 1998, p. 57). By documenting and reporting each step in the research process, giving the rationale for decisions that are made, and confirming that the data accurately reflect the narrators' perception of the experience, the researcher can provide the markers or signposts that others might follow to continue or extend the study. These are solid indicators of replicability.

Objectivity, lack of bias

In any research project, whatever the methodology, the potential for subjectivity exists, and in interview-based research, a personal inclination to a particular point of view could introduce a bias that would make the research of dubious worth. Instead of asserting that it is possible to achieve a distanced objectivity, it seems more productive to proceed from a position of reflection, candor, and disclosure.

Accounting for subjectivity involves reconciling interview data against other reports and data sources, attending to inconsistencies that may emerge, and appraising the potential for distortion that may enter at the analysis and interpretation stage. Having background knowledge of issues or prior knowledge of the topic being studied brings a potential for greater perceptivity, but it also introduces a potential liability, namely, a greater likelihood for bias and a failure to notice subtleties or disconfirming evidence. The ability to see something is influenced by what we think we should look for, and the result can be inattention to other details that emerge. We develop a language to communicate our experience, and that very language influences our perception and results in our attention to qualities that have particular value to us (Eisner, 1998). With a research topic that has enough personal appeal to warrant months of study, it is important to recognize that, as Eisner noted, "A way of seeing is also a way of not seeing. . . . What we see is frequently influenced by what we know" (Eisner, 1998, p. 67).

"The real aim of a life-history sociologist . . . should be to reveal sources of bias, rather than to pretend they can be nullified, for instance by a distanced researcher without feelings" (Thompson, 2000, p. 137). It is possible to minimize the negative impacts of prior knowledge or subjective experience by facing this challenge directly. While it is true that the hazards of unbridled subjectivity must be avoided, it is also true that the quality of an interview can depend on rapport that comes from a subjective knowing. Subjectivity has its virtues, a point that the method of educational connoisseurship and criticism has always recognized (see Eisner, 1998). Eisner employed the term the *connoisseur* to describe one who has deep knowledge about a subject and who can best use that knowledge judiciously in interpreting meanings and nuances that might be missed by others less informed. Adept researchers capitalize on their prior knowledge and experience, turning what might have been a drawback into an asset. They also check for meaning and confirm that they are correctly interpreting what their narrator is saying without prejudging based upon their prior knowing.

Monitoring personal reactions throughout the investigation can help a researcher become conscious of any subjective lens that is being activated (Peshkin, 1988). Noteworthy reactions during an interview, such as avid agreement or disagreement with a participant's narrative, extended moments of inattention, or a strong connection that triggers personal memories, can signal the possibility of subjectivity being aroused. These factors need to be observed, reflected on, and disclosed in reporting the research. The reason for this disclosure is not to compose a personal testimony of deeply held opinions, but to acknowledge that those opinions do exist and to explain the steps that were taken to minimize their negative effect. This requires forethought in designing the study and vigilance in selecting study participants, conducting the interviews, reporting the data, and drawing conclusions from the data. It certainly requires that the potential for bias be disclosed to the reader.

Utility

At the beginning of this chapter, I emphasized that the purpose of the research shapes its design, so considering the utility of interview research is really a question of the purposes that it can serve.

I especially admire Seidman's (2006) straightforward take on this matter:

> The purpose of in-depth interviewing is not to get answers to questions, nor to test hypotheses, and not to "evaluate" as the term is normally used. . . . At the root of in-depth interviewing is an interest in understanding the experience of other people and the meaning they make of that experience. (p. 9)

The usefulness of an interview-based study, therefore, relates to the purposes that can be served by sharing this expanded understanding of the events and circumstances of the lives of individuals so that others can better comprehend their actions, decisions, responses, perceptions, and beliefs. By looking beneath the observable phenomena, interview research should reveal the underlying significance.

With the focus on deepening understanding by accessing the memories of a small number of individuals within a given context, it is safe to say that the results of interview research may not be widely generalizable, at least not in the standard use of that term. Yet such studies do produce findings that have implications for other settings and situations. They tell much about human nature and the way people interact with their world, face their challenges, and make meaning from their lives. Interview research provides a perspective that can inform others about the effects of actions and decisions on the lives of individuals within the context being studied. It tells about the way people live their life and sheds light on matters of significance, bringing events into clarity in human terms. As with any research investigation, utility needs to be assessed based on whether or not the research contributes something of value to the field of study, the appropriateness of the sample of informants with regard to the question and purpose of the study, the thoroughness of the analysis, the quality of the findings, and the situations to which those findings may be reasonably applied.

WHAT'S NEXT?

Deep involvement in researching human experience poses multiple challenges, extending from the personal to the academic. Your selection of a topic, your own preferences in managing the research, your ability to produce an investigation that measures up to the standards for research are all considerations for the journey. In-depth interviewing for an ethical purpose offers to the interested and predisposed researcher an opportunity to explore uncertain territory—perhaps facing a few dragons of doubt and bewilderment along the way—but ultimately with the potential for contributing an awareness and appreciation for experience that can open worlds of discovery for others. It is this spirit of inquiry that is vital to a qualitative pursuit using in-depth interviews.

In the following chapter, you will have an opportunity to consider the principles that govern the conduct of ethically sound investigations. All research needs to be envisioned, planned, conducted, and reported with a commitment to the highest of ethical standards. This is especially true when research seeks to investigate the way that individuals experience and internalize life events. To understand a situation, it is necessary to gain access to the way people perceive it, not just analyze what appears to have happened from an external vantage point. Actions may have been motivated or designed for one purpose, yet it is the way those actions are perceived and internalized that truly inform the question of the ultimate consequence or meaning that people take from the experience. This requires asking people to look on the inside for their understanding of what happened to them on the outside. Researchers encouraging this type of reflection must ensure that their motives are ethical and that they have designed an investigation that safeguards their participants' well-being. Chapter 3 will help you be able to do just that.

Chapter 3

Ethical Research Practice—Doing What's Right

Chapter Topics:

- regulations governing human subject research
- ethical principles
- institutional review boards
- research among diverse or marginalized populations

When you ask questions that set individuals to searching their memories with intention, it's not always clear what the outcome will be. As a conscientious researcher, you need to keep sight of the possibility that what transpires in an interview could put your narrator at some discomfort, disadvantage, or jeopardy. While perhaps not as risky as biomedical research or pharmaceutical testing, social science and educational investigations are not without a potential for harm. Agencies and institutions receiving federal funds are required by law to maintain protective oversight of proposed research investigations that involve living human beings. Subsumed within these mandates is the simple matter of ethical behavior to govern research practice.

Ethics is the branch of philosophy dedicated to the study of moral theory and behavioral norms ranging from absolutism to relativism. Within this broad field lies much room for discourse and contemplation. For this book, I would like to focus on a fairly straightforward view of ethical behavior, and that is the assertion that an ethical researcher is motivated by ethical purposes; is guided by principles of fairness and equitability; and is committed to do no harm to the individuals in the study, to faithfully present findings, and to fulfill both the letter and the intent of legal requirements for research conduct. Since education and social science research involves the study of people and society, let's start with the fundamental responsibility to protect the safety and welfare of study participants.

REGULATIONS AND RESPONSIBILITIES

Researchers must protect the rights and safeguard the well-being of the individuals and groups who participate in their investigations. Extreme abuses of this basic principle in the past have led to binding regulations for research practice. While these legal concerns weigh heavily on the researcher who is working within any organization that receives federal funding, a moral code of behavior extends well beyond what is legally required. "Central to the responsive interviewing model is the importance of obtaining rich data in ways that do not harm those being studied" (Rubin & Rubin, 2005, p. 97).

The following discussion is not intended as a comprehensive analysis of the regulations governing human subject research nor does it claim to offer specific guidance or advice for the conduct of such investigation. My goal here is to introduce some of the issues of ethical practice for you to consider as you begin your research.

Reasons behind the rules

The simple mandate of *do no harm* is such a fundamental tenet for moral and humane behavior that it is difficult to comprehend some of the violations that have been committed under the guise of sanctioned research. As early as 1802, Dr. Thomas Percival advocated peer review for proposals to conduct medical research involving humans (Cleary, 1987), and yet it wasn't until after World War II, almost 150 years later, that actions were taken to eliminate or at least reduce the potential for research abuse.

During the war, Nazi scientists and medical practitioners conducted cruel and barbaric experiments on prisoners who were helpless to defend themselves. Forced organ transplants, exposure to lethal pathogens, mutilation, sterilizations, and other unconscionable acts revealed a complete disregard for basic human rights and welfare. Following the war, those who perpetrated the experiments were brought before an international military tribunal at Nuremberg, Germany; many were convicted and executed for their crimes. In an attempt to prevent such egregious abuse in the future, the Nuremberg tribunal generated a list of conditions that should be met in conducting experiments involving human beings. These standards, which became known as the Nuremberg Code, required that subjects of experimentation give their voluntary and informed consent and that any such experiment should offer potential societal benefits without causing participants unwarranted pain or injury (Nuremberg Code, 1949).

The Nuremberg Code was echoed in the Declaration of Human Rights and affirmed in principle by the original 51 members of the Charter of the United Nations. At that time, there was no mechanism for implementing the provisions of the code, either in the United States or in other countries. In 1953, the Clinical Center of the National Institutes of Health created the first U.S. federal policy for the protection of human subjects (U.S. Department of Health and Human Services, 2004). These actions marked the beginning of the research review process that is now integral to the human subject protections in the United States. The principles delineated in the Nuremberg Code and in subsequent charters were successful in reducing the severity of research violations, but they did not put an end to unethical practice altogether. As a result, experimentation without proper regard for human protection continued.

It should be noted that human subject abuses were not limited to investigations conducted in other nations. In the United States, among the most notable violations was the Tuskegee syphilis experiment, which began in the 1930s and continued for 40 years so that researchers could track the long-range effects of syphilis on nearly 400 disadvantaged African American men. These men had agreed to participate, but they had not been fully informed about the risks they faced or the design and intent of the study (Seidman, 2006). Perhaps the most damning aspect of the Tuskegee experiment was that researchers withheld antibiotics from the study participants even after viable treatments became available.

When the facts of the Tuskegee study were made public, the U.S. Congress called for an extensive investigation and in 1974 established the National Commission for the Protection of Human Subjects of Biomedical and Behavioral Research. This commission was given the task of framing ethical principles to undergird the conduct of biomedical and behavioral research and developing guidelines to ensure that research measures up to these principles. Several years later, the commission issued *The Belmont Report: Ethical Principles and Guidelines for the Protection of Human Subjects of Research* (1979). Widely known by its shortened title, *The Belmont Report* defines research as an activity that states an objective, specifies a set of procedures, tests a hypothesis, allows conclusions to be drawn, and contributes to generalizable knowledge. The report specifies three ethical principles for the protection of human research subjects:

1. Respect for persons, including informed consent, right to privacy, confidentiality, and safeguards for those of limited autonomy;
2. Beneficence, namely to do no harm, both maximizing benefits and minimizing potential for risk; and
3. Justice, including fairness and equity in selecting study subjects and sharing the benefits of the research with all who participated.

Respect for persons includes matters of physical, mental, and emotional well-being; personal safety; dignity; confidentiality of records; and assurances of informed, voluntary participation. Obtaining informed consent, a critical element of this principle, is the process of informing potential research participants about the study, its purpose, the procedures that are involved, the risks as well as anticipated benefits that might come from the research, and a signed statement confirming that they know their involvement is voluntary and that they may withdraw at any point in the study at no penalty or loss. Researchers are responsible for providing this information in language that participants can fully comprehend. "Because the subject's ability to understand is a function of intelligence, rationality, maturity and language, it is necessary to adapt the presentation of the information to the subject's capacities. Investigators are responsible for ascertaining that the subject has comprehended the information" (National Commission for the Protection of Human Subjects, 1979, Part C.1, para. 8). This criterion necessitates special considerations for participants who are of diminished capacity, such as young children, the mentally disabled, or patients with a terminal health condition. All potential participants, regardless of capacity, must be given the opportunity to consent or decline involvement, to the extent to which they are able. If the individual is of limited capacity, the consent to participate must be granted by a responsible party charged with protecting that person's best interest; in most cases this means a parent or legal guardian. At the heart of the requirement for informed consent is that participation in the investigation is truly voluntary, with full awareness of risk, and without excessive pressure or promises of unlikely reward.

Beneficence requires a thorough assessment of risks and benefits, a criterion that necessitates a careful analysis of the possibility that harm might come to participants as well as a realistic appraisal of potential benefits that might be gained. Application of this provision requires that researchers assess the possible impacts of the study on the subjects, their families, and society at large. This necessitates a consideration of risks versus benefits to the individual as well as the potential societal benefits that could be gained from the research. Risks, whether of a psychological, physical, legal, social, economic, or more general nature, include a probability that participants would experience any harm, either during the study or after it is completed. This requires that researchers consider immediate risks as well as long-term implications. A potential health risk during a medical study, for example, may be clearly recognizable; but the researcher also needs to consider risks of emotional, psychological, or physical consequences that might not show up until years after the study has been completed. Investigators must be diligent in considering the potential for harm. Benefits, those positive outcomes that may result from a study, are not expressed in terms of numerical probabilities. Instead, they are generally contrasted with harm, in a *risk/benefit* format, to reveal the magnitude of potential harm and the anticipation of likely benefit.

Justice speaks to the matter of equity in the selection of study participants as well as in the equitable distribution of benefits from the research itself. Selection of participants must be determined through procedures that are fair and just. Research investigations should not require that members of any particular group or social class bear all of the risks or burdens or that any particular group exclusively reap the benefits. The issue of social justice is a special concern in research that involves marginalized populations or vulnerable individuals. Certain groups, such as immigrants (regardless of citizenship status), racial minorities, the aged or infirm, as well as those who are mentally disabled, economically impoverished, incarcerated, or institutionalized, are particularly vulnerable. Research among these populations must be designed with heightened sensitivity to safeguarding of subjects' rights and welfare.

The Belmont Report is the philosophical basis for the Code of Federal Regulations, Title 45 Part 46, Protection of Human Subjects (commonly referred to as 45 CFR Part 46), which is considered the *common rule* that aligns all human subject research conducted through U.S. federal agencies and entities receiving federal funds. The regulations within this rule apply to all research involving living people, but they do not codify universals to govern decisions about ethical correctness of proposed research. Instead, they delineate specific criteria that must be met and then decentralize the governmental role in approving research by requiring that institutions, agencies, and organizations receiving federal funds create a local institutional review board (IRB) to ensure that proposed research is ethically sound. This framework is designed for the purpose of protecting the rights and well-being of all research participants.

The Nuremberg Code, *The Belmont Report*, and 45 CFR Part 46 are all available online through the U.S. Department of Health and Human Services, Office for Human Research Protection, at http://www.hhs.gov/ohrp/. Additionally, these documents may be accessed through the Office of Human Subject Research of the National Institutes of Health, at http://ohsr.od.nih.gov/guidelines/index.html. A number of revisions in the rules have been made over the years, so check the official Web sites to be certain that you have the most up-to-date edition of the mandates. If you have difficulty locating any of this material, you can search by title through your Internet search engine. In addition, you should be able to locate copies through your institution or organization's office that oversees human subject research.

For more information about these regulations, you might consult the documents previously discussed or, for a broader perspective, consult *Protecting Participants and Facilitating Social and Behavioral Research*, a publication produced by members of the Panel on Institutional Review Boards, Surveys, and Social Science Research for the National Research Council of the National Academies (Citro, Ilgen, & Marrett, 2003). This work is available online through The National Academies Press at http://www.nap.edu. You might also want to read any of a number of reputable histories of the Nuremberg trials or the Tuskegee syphilis study, or simply conduct an academic-database search for articles on violations of human subjects research ethics. The extent of the abuses that federal regulations were designed to prevent brings the task of applying for IRB approval into some perspective. While I realize that perusing these legal documents is not a requirement for doing research, it is an excellent place to ground your investigation. Knowing the basis for the regulations can help you plan your study and meet the ethical standards for research.

The work of the IRB

Charged with protecting research subjects and ensuring that research complies with principles outlined in *The Belmont Report* and the specific protections provided in 45 CFR Part 46, universities and organizations sponsoring research have established critical review boards to oversee the approval process. While each institution's review board is locally formed, all must be composed of at least five members who have expertise and competence in evaluating research plans, including at least one member with specific familiarity in the field of study under consideration. IRBs must include at least one community representative (an individual neither affiliated with the institution nor connected to the researcher). The exact name of these IRBs varies by institution—for example, human subjects research panels, office of sponsored programs, or research ethics review boards. However, all share the same mandate: Assure compliance with regulations for the protection and ethical treatment of human research subjects.

Federal regulations are insufficient to cover every possible permutation or complexity of all possible research; instead, they contribute the foundation and framework that IRBs use in considering all applications to conduct research. Among the first considerations of an IRB are the questions *Is it research? Are human subjects involved?* and *Do the regulations apply?*

For the purposes of IRB consideration, *research* is defined as "a systematic investigation, including research development, testing and evaluation, designed to develop or contribute to generalizable knowledge" (45 CFR Part 46.102[d]). A *human subject* is a living person about whom a researcher obtains private, identifiable information or data through communication, observation, interpersonal contact, or interventions such as testing. Some types of research investigations are eligible for exemption from these regulations, and some are eligible for an expedited review process. However, it is the IRB that makes these determinations, not the researcher.

Before you will be allowed to conduct research through an institution that receives any federal funds, you will need to demonstrate that you have completed training that certifies your understanding of protective regulations governing human subjects research. While some universities continue to hold their own requisite training program, many rely on the online training program of the National Institutes of Health and Collaborative IRB Training Initiative. In 2008, more than 850 organizations and institutions were employing this program as part of their ethics education programs (CITI, 2008). Your IRB office will be able to guide you in completing the certification process that is required for your specific institution or organization.

Since IRBs are locally constituted, each determines the specific process and required application form for use in that setting. In general, though, IRBs require researchers to submit a detailed and extensive application describing the proposed study, the researcher's qualifications that demonstrate capacity to complete the study, the methodology to be employed, the proposed study sample, how participants will be selected, the risks and benefits, how informed consent will be acquired, and the precautions for protecting confidentiality as well as the safety and well-being of all participants including, when appropriate, the safety of the researcher. The requirement that informed consent be obtained prior to the research is intended as a way to ensure that all participants have been informed about the nature of the investigation, their rights to decline participation, and the possible risks and benefits associated with being involved in the study. Only those individuals who document that their consent to participate is based on this requisite information can be allowed to take part in research. While some research employs surveys or other tools that allow for anonymity of the participants, the nature of interview research means that the respondent is not anonymous to the interviewer. Thus, a commitment to protect confidentiality, and not guarantee anonymity, is the standard.

Title 45 CFR Part 46.116(a) requires that an informed consent form include the following eight elements:

1. A statement that the study is research, the purposes of the research, the duration of the participant's involvement, a description of the procedures, and specification of any procedures that are experimental;
2. A description of possible risks or discomforts;
3. A description of possible benefits;
4. Disclosure of any alternative procedures;
5. A statement describing what will be done to protect confidentiality of records;
6. A statement of any compensation or recompense;
7. Contact information for individuals to consult regarding any questions about the research, participant's rights, and what to do if a problem develops; and
8. A statement specifying that the participation is voluntary and that the subject has the right to withdraw from the research at no penalty or loss.

In addition to these required elements, specific circumstances may necessitate including one or more of the following in an IRB application:

1. A statement of any risks that may be encountered if the participant is, or should become, pregnant;
2. Any circumstances under which the participant's involvement may need to be terminated without his or her consent;

3. Any expenses that the participant may be expected to bear as a result of the research;
4. Consequences to the participant if he or she decides to withdraw from the study and the procedures that have been set up for an orderly withdrawal should it be necessary;
5. A statement that any significant new findings from the research that relates to the participant's willingness to continue in the research sample will be provided to the participant; and
6. The approximate number of participants who will be included in the study.

IRBs meet regularly to consider applications for approval to conduct research under the auspices of that institution or organization. In fulfilling this role, IRBs must feel confident that the research will not endanger the participants. IRBs also attempt to minimize the potential for any liability claim or litigation against the institution.

While I don't want to turn this into an essay on IRBs, I do need to acknowledge that many education and social science researchers feel that the risks in their discipline are not commensurate with the risks in biomedical research that IRBs were originally designed to address (see Denzin & Lincoln, 2005). These researchers are concerned that the current movement toward methodological conservatism that emphasizes "scientifically based" or "evidence-based" research may pose a threat to qualitative research (Lincoln, 2005). Members of IRBs have been known to deny research applications because of their own particular biases for research methodology or their own anxieties about the potential for risk, even when reasonable and adequate safeguards are in place (see Cleary, 1987; Fitch, 2005; Marshall, 2003). IRBs are simply committees of individuals, and thus, it is not surprising that an individual on an IRB might bring personal perspectives and personal fears to bear in making decisions that may hit a little close to home. It helps to find out as much as possible about the workings of the IRB at your institution before you begin the application stage, perhaps consulting with a member on the board in advance of drafting your proposal. In my own case, it was with the support and problem solving by my dissertation committee along with direct communication with a member of the board that I was able to gain approval to conduct what was considered sensitive research among fellow parents in my community.

Many oral historians argue that their work should not be subject to IRB approval since historical research is not intended to be *generalizable* and since the basic purpose of the work (i.e., to contribute to the historical record) is in conflict with the government requirements for anonymity/confidentiality. In 2003, the Oral History Association announced seeming consensus regarding IRB oversight:

> The U.S. Office for Human Research Protection (OHRP), part of the Department of Health and Human Services (HHS), working in conjunction with the American Historical Association and the Oral History Association, has determined that oral history interviewing projects in general do not involve the type of research defined by HHS regulations and are therefore excluded from Institutional Review Board oversight. (Oral History Association, 2003, para. 1)

This clarification in interpretation did not quite settle the argument about whether IRBs should oversee oral history work, however, and controversy at the local level remains. "While some universities have agreed that federal regulations were never intended to cover oral history research, many other Institutional Review Boards are holding fast to rules that include oral history under human subject research, despite recent communications to the contrary from the concerned federal government office" (Townsend & Belli, 2004, para. 1). The best advice is to assume that, if you are conducting research involving living human beings, your work is subject to IRB approval.

Informing and inviting consent

Given this brief review of the regulations for human subject protection, the question now becomes, what does it mean for interview-based research? Interviewers do not conduct experiments involving risky medical procedures or questionable pharmaceutical testing. However, it must be acknowledged that, even in interviewing, there exists a potential for harm and that precautions are prudent.

Establishing an environment of respect and trust is a key contributor to participant protection. The participants in your research need to know that you will treat them fairly; that you will do all in your power to guard their identity from disclosure; that you have truthfully described your study, its intent, and all reasonable potential for risks that they may encounter; and that you have not misrepresented any potential for benefit. They must fully understand where and how the information that they share with you will be used and for what audience. You must make sure that they understand the voluntary nature of their participation and that they retain the right to decline to answer any question and may withdraw from your study at any point, even if you have already completed the interviews. You also need their signed consent to audiotape or videotape the interview sessions. I have provided a sample informed consent form in Appendix B-1 to help you get started, though you need to check with your IRB to make sure of their exact requirements.

The details spelled out on the informed consent form establish the areas you will address as you ask potential study participants to agree to join your study. In addition to these fundamentals, you need to be clear that there are limits to what you can and cannot do to protect the privileged information that they are sharing with you. Remind them that if, for any reason, your research is subpoenaed, you might be forced to comply with a court order for access. If, at any time, you see the conversation moving in a sensitive or problematic direction, consider turning off the tape recorder to repeat these cautions as well as to advise the narrator against making potentially defamatory or libelous remarks. If a study participant discloses situations involving child abuse or neglect or any situation in which you feel that the individual poses a danger to himself or to others, you need to break the promise of confidentiality and act in a responsible manner to protect those involved. If your research does produce disclosures that lead you to question how to legally and ethically respond, discuss your dilemma with your advisor or legal counsel.

With all of these caveats and cautions, it might seem unlikely that anyone would agree to be interviewed, but people will agree. They just need to be given complete information as well as the time to ask questions and become comfortable with what you are proposing and why. If narrators seem unsure, it can be helpful to remind them that they may decline to answer any question, may terminate an interview, and may withdraw altogether. Asking for a signature at the beginning of an interview may seem to put the narrators at a disadvantage, because they are in effect giving you permission to use their responses before they even know what the questions will be. Nevertheless, in research that is governed by IRB oversight, you must have their signed consent before you can begin to interview them. Since an in-depth interview project will usually involve a minimum of three sessions, narrators will be able to assess their comfort over a period of time and withdraw if they feel anxious or uncomfortable. However, at no time should you pressure an individual to join your study or to continue if it is clear that he or she is feeling anxious or uncertain about participating.

While informed consent is intended to protect all narrators, both the powerful and the powerless, the protection of marginalized or disadvantaged individuals requires special vigilance. Full understanding of rights and privileges must be confirmed, and a potential narrator's lack of familiarity with formal language on a consent form could limit the full comprehension of what is being agreed to. Similarly, when faced with signing a legal document, some may decline simply because of uncertainty about larger implications beyond your research study. If you are conducting research into a disadvantaged community or within a group that speaks a different primary language than you do, it might help to connect to a member of that group to help you bridge the gap in communicating with potential narrators. Ask for help in drafting the informed consent document and guidance in how to introduce and explain your work to members of that group. Equity in research requires that efforts be made to learn across all populations. However, ethics demands that research participants fully understand and consent to participating in the research.

If you are asking your narrators for permission to interview them as a part of your dissertation research with no plan for further publication or presentation, then your informed consent form should state that as your intention. However, by signing this form, your narrators give you permission only to use their interviews for the purpose and distribution that you specify. The caution here is that if you do find subsequent opportunities to publish or present on your research then you might need to return to your narrators for their further consent. Ask your advisor or someone from your local IRB office for examples of wording that they feel is acceptable for protecting the narrators' rights while allowing you to use the research as presented in your dissertation elsewhere.

You also need to remember that an IRB approval gives you permission to work with the data only during the period for which the approval is in effect, generally one year. If you want to make changes to your research design within the approval period, you will need to apply for IRB permission to do so. If you are not able to complete your data analysis in that period of time, or if you later decide that you want to return to the data for additional analysis after you've completed the approved research, you need to apply for a continuation or a renewal of the IRB approval. IRB approvals also require that you maintain all data related to the research for a period of three years after the completion of the study.

The proverbial elephant in the room regarding the equitability of research cannot be ignored. The potential exploitation of another's time and voice for the advantage of the researcher is a troubling matter. Certainly, if the researcher anticipates considerable financial reward, then narrators may have a right to an equitable share. However, "there is a more basic question of research for whom, by whom, and to what end. Research is often done by people in relative positions of power in the guise of reform. All too often the only interests served are the researcher's personal advancement" (Seidman, 2006, p. 13). As discussed in Chapter 2, it is important to clarify your purpose for conducting the research so that you will be able to state with conviction why you are doing what you are doing. If you are researching exclusively for personal gain (as in the pursuit of your graduate degree), then I encourage you to consider how your research might also benefit those whose lives and experiences you are studying.

BEYOND THE IRB

The ethics of research extends to equity in selecting narrators for your study, fairness in their treatment, and the integrity with which you handle the information they give you. As explained in *The Belmont Report*,

> . . . selection of research subjects needs to be scrutinized in order to determine whether some classes (e.g., welfare patients, particular racial and ethnic minorities, or persons confined to institutions) are being systematically selected simply because of their easy availability, their compromised position, or their manipulability, rather than for reasons directly related to the problem being studied. (The National Commission for the Protection of Human Subjects, 1979, Part B)

Your IRB will closely examine your application for this sort of equity and social justice. However, another type of research equity also demands attention, and that's the equity relating to the conduct of research on topics that may be of special concern to marginalized populations but, perhaps, of limited interest to the public at large. Being committed to research with an ethical purpose requires a consideration of topics that may be challenging to access yet vital to explore. Recognize the diversity of experience, and then consider researching issues that allow you to explore diverse perspectives that can inform others about unseen situations. As an in-depth interviewer, you have an opportunity to make the problems of your narrators "more visible, raising the level of public discussion; at best, you can come up with policy proposals that will ameliorate your interviewees' problems and try to get those solutions adopted" (Rubin & Rubin, 2005, p. 101). The stories of life experience within a group that is underrepresented in research can help to create awareness and bring attention to matters that may have been overlooked or avoided. "The real service I think we provide to communities, movements, or individuals, is to amplify their voices by taking them outside, to break their sense of isolation and powerlessness by allowing their disclosure to reach other people and communities" (Portelli, 1997, p. 69).

Regardless of the topic you select or the participants you engage in the pursuit of answers, you are responsible for achieving an ethics of accuracy. Narrators have a right to expect that their meaning will be accurately represented and not put to any use other than what they agree to. "Events do not speak for themselves, and meaning is not a natural occurrence. Context and perspective help give meaning" (Schuman, 1982, p. 2). Researchers have a responsibility to communicate an experience through the narrator's eyes, providing enough background so that the reader can understand the meaning and significance of those disclosures. This portrayal should be respectful, without tones of sarcasm, ridicule, or authority. While it can be an empowering experience for an individual to be asked to express opinions and life stories, you need to remember that you aren't giving a voice to this individual; you are merely providing a medium for the voice to be heard by others. You are there to listen while narrators give voice to their own experience. In Chapter 7, you will learn some helpful strategies for ensuring that the voice that is heard in the interview is the one that is shared in your work.

Matters of voice and ownership lead to questions of the fairness in "taking" stories from narrators while leaving them with little to show for their efforts. This is best expressed in the oft-quoted article by Patai (1988), who questioned whether scholarly research isn't inherently exploitive, since the researcher benefits to a greater degree than does the individual whose life is researched (Seidman, 2006). At an absolute minimum, researchers have a responsibility to dutifully and accurately represent those stories and to appropriately acknowledge the contributions that the narrators have made to the research. One safeguard to ensure accurate representation is "to provide [interviewees] with the rights to look over their interviews, edit them, and examine the final manuscript before publication. . . . You need not promise your interviewees a veto over what you conclude, but you should assure your interviewees that you will be fair and proceed in a balanced way" (Rubin & Rubin, 2005, p. 10). The conclusions are yours, but the meaning that is the basis for those conclusions originates with them.

Issues of ownership and copyright are rightly considered within the realm of ethical practice. Legal statutes define rights to copyright as well as ownership of the audiotapes and transcribed interview data. While the contents of the interview come into being through the collaborative interaction of the researcher who asks the questions and the narrator who responds to them, the life story being told belongs to the individual who lived the experience. As Yow explains (2005), under the terms of the U.S. Copyright Act of 1976 and the Digital Millennium Copyright Act of 1998, the moment an interview ends and the recorder is turned off, the tape that has been produced belongs to the narrator. Since the narrator has rights as the author of its contents, researchers need to include, on the informed consent form, a statement by which the narrator grants permission (a legal release) for the tape's use as specified for the research.

As you design your research, you must specify how you will dispose of the interview recordings. For purposes of educational and social science research, tapes are rarely archived due to issues of confidentiality and protection of research participants. If, however, the interviews have been conducted for the purpose of enriching an oral history collection, you need to determine archival details before undertaking the research. In this latter case, the narrator needs to sign a deed of transfer to the archive wherein the tapes are to be housed, with terms specifying their use, by whom, and at what point in time. In many cases, narrators request that tapes not be released until after their death, and these conditions must be clearly spelled out in any such agreement.

What's Next?

Interviewing as research requires meeting standards for ethical practice while providing direct access to the unseen world of human perception and memory. Additional resources to help you become familiar with these standards and regulations governing human subjects research are provided in Appendix A-1. These considerations, along with viable research purpose and questions, are fundamental to the design of any interview-based research investigation.

Chapter 4 builds on these common practices of interview research to introduce the particular traditions that contribute to the design of a gateway study. These traditions include elements and epistemology characteristic of *educational connoisseurship and criticism*, a qualitative practice developed by Elliot Eisner for the study of schools; an open-ended approach to in-depth interviewing characteristic of *oral history*; a form of data reduction derived from *poetic transcription*; and a means for confirming interpretive accuracy comparable to a *member check*. Following a consideration of these elements, Chapter 5 begins the discussion of how to adapt these strategies for research using the gateway model.

CHAPTER 4

FOUNDATIONS FOR A GATEWAY

CHAPTER TOPICS:

- educational criticism
- oral history
- poetic display
- member check
- implications for research

Previous chapters considered aspects of research that are characteristic of all qualitative investigations that use interviews as the primary means of data collection. A clearly framed guiding purpose, critical research questions, compatible researcher disposition and skill, and ethical conduct are requisites for the research, regardless of the particular methodology being employed. After these first factors have been considered, the path will lead some researchers to commit to in-depth interview research. For those investigators, the gateway approach offers a model for crossing the boundary of cause and consequence that separates individuals who have lived an experience from others who have not.

The gateway approach has been recognized as an innovation in qualitative research, and I have wondered exactly what makes it so. It isn't innovative because researchers have never before conducted in-depth interviews and reported data in the form of an excerpted display—they have. Nor is it innovative because researchers don't document targeted situations or events—they do. Further, it isn't new because investigators haven't carefully crafted precautions for dealing with issues of subjectivity—they have. And it certainly isn't new because its arts-based, literary style of disclosure is a novel invention—it isn't.

I have concluded that what distinguishes the approach is that it offers a creative means of connection, so that the researcher, the narrator, and the reader all expand their understanding, not only of external circumstances but also of emotionally charged responses and perceptions. The approach, which capitalizes on human expression in ways that bring life to rigorous academic research, is grounded in the awareness that an experience does more than trigger cognitive processing. Along with the attempt to make meaning and understand from a mental frame, humans relate to an experience on many different levels—psychological, emotional, social, physical, philosophical, and so on. By honoring all dimensions of experience, the gateway model provides access into sometimes confusing and dark places and addresses the internal component that is often stripped out or neutralized in statistically based research. It offers an integrated process that allows a researcher to capture the essence of an experience on as many levels as are experienced by an individual.

A fundamental principle of gateway research is that what is learned should reflect the narrators' understanding and be communicated in a way that can inform action that serves an ethical purpose. With the goal of equity guiding the process, it provides a way of knowing that offers a way of helping. With apologies to James Baldwin, "You write something to change the world. . . . The world changes according to the way people see it, and if you can alter, even by a millimeter, the way people look at reality, then you can change it" (Pipher, 2006).

WHAT IS GATEWAY RESEARCH?

The gateway approach is a narrator-centered model for interview research into the "reality" of a life experience through the perspective of others. An open-ended style of questioning from the tradition of oral history invites narrators to share stories of their experiences and to consider the impacts of those experiences on the many dimensions of their life. By distilling transcripts of those interviews in a way that communicates the essence of the experience, it is possible to provide a clearer expression of the thoughts and feelings of the narrators, thus keeping people *present* in the research, not just *represented* in summaries or paraphrased profiles.

Gateway researchers build background knowledge related to their topic and a specific awareness of the culture and language of individuals in that unique setting. Attention to nuances of language enables connection, and by respecting and accurately interpreting what narrators express, it is possible to cross borders of understanding so that those outside of a situation can learn from the experiences and responses of those inside who are willing to share their stories. At the same time, the approach provides a way for those inside to reach across and communicate with those outside, sharing what it meant to have that experience and creating a community of deeper understanding about it. Contributors to studies using the approach find that their participation also offers a pathway for personal reflection, and in this regard it may serve as a gateway to deeper self-understanding.

I'd like to share with you one of my own experiences that taught me the importance of this research perspective. Years ago, I agreed to participate in a research investigation and was interviewed on three separate occasions for over an hour each time. When the interviewer asked me to read the transcript for accuracy, the most common form of member check, I promptly reviewed the document and returned it with only a few changes. When the final report was released, I was stunned to see that I had become invisible. Interview data for all 12 participants in the study had been reduced into a quantified coding that stripped our narratives of contextual meaning. In addition, the report only enumerated individual responses, so instead of qualities of actual human experience, the report provided quantified generalizations (*some*, *all*, *2 of 12*, etc.). I felt that my voice and the voices of others in the study had been silenced, our personal interpretations of our own experiences negated. That encounter on the other side of the microphone made me committed to preserving each narrator's voice and meaning in my research.

FRAMES FOR THE GATEWAY

Different research methodologies bring different ways of accessing data, different relationships to the data, different forms of analysis, and different epistemologies or ways of knowing what is known and believed. When I started looking for a viable way to conduct research into the Columbine tragedy and to discern meaning that could be shared with others, I could not find a traditional method that would serve my purpose. I did find strategies that I could adapt and modify to meet my needs, but I was not comfortable launching out on my own. I wanted to be certain that the use of strategies I was considering was defensible and supported by the research traditions. I needed to know, for example, that what I was proposing was not violating paradigms for oral history or educational criticism because I did not want to misrepresent or misuse the terminology from those existing practices. Similarly, I wanted to confirm the legitimacy of my application of poetic transcription and member checking. Words have power, and because I respected the language of research and the language of others who had developed the processes I was adapting for my own purpose, I felt it was essential to research the research traditions.

You may not feel compelled to do likewise, but the following overview may help you understand the functionality of the gateway approach. With this background, you can better recognize the philosophical underpinnings as well as the purpose, process, and possible applications of the gateway model.

Educational connoisseurship and criticism

Even though the gateway approach is a form of in-depth interview research, many of its practices have been influenced by tenets of educational connoisseurship and criticism, a qualitative research method initiated by Elliot Eisner that relies primarily on informed observation for data collection. Eisner's method (hereafter referred to as *educational criticism*) emphasizes the perception of qualities, the interpretation of significance of what is seen, and the giving of public form to the content of consciousness (Eisner, 1998). Eisner envisions the method as a means to learn about schools and classrooms, affording the researcher an opportunity to view and consider meaning from instructional practice with a thoughtful and informed eye. Its purpose is

. . . [to] contribute to the enhancement of the educational process and through it to the educational enhancement of students. . . . It is not a value-neutral descriptive vehicle concerned only with something called disinterested knowledge but rather is concerned with understanding for educational improvement. (Eisner, 1998, p. 114)

Achieving this high purpose requires that the researcher be able to discern and discriminate among complex and subtle qualities that are observed and to understand the conditions that precipitate the development of these qualities. Antecedent knowledge of this type is foundational to *connoisseurship*, which, as Eisner describes it, is the art of appreciation. That which we appreciate—art, wine, or classrooms—become things we want to know more about (Eisner, 1998). As we learn more, we can eventually become connoisseurs, but unless what is discerned is shared, connoisseurship serves a limited purpose. This is where criticism comes in, for *criticism* is the art of disclosure that communicates what is discerned by the connoisseur in ways that make the significance of what has been seen apparent to others.

I must admit that when I first approached the Columbine study, I was troubled about the connotations carried by the term *connoisseur*, which seemed lofty and hierarchical. While I saw great value in the premise that successful research requires deep knowledge of the topic and refined appreciation of the qualities contributing to its meaning, I was not comfortable with the prospect of becoming a connoisseur of school shootings. The implications were somehow disturbing, but ultimately I realized that the term simply meant that I needed to develop background knowledge and appreciation for the multiple levels of nuance and complexity to prepare me to make fine-tuned and informed discernments. This perspective made sense to me, and the deeper I looked into Eisner's method, the more applicability I saw.

As an art educator, Eisner approached educational inquiry from the model of criticism employed by art critics, whose task is "to function as a midwife to perception, to so talk about the qualities constituting the work of art that others, lacking the critic's connoisseurship, will be able to perceive the work more comprehensively" (Eisner, 2002, p. 213). As Eisner points out, it is antecedent knowledge that influences our ability to perceive and interpret events and experience. It shapes our awareness of and appreciation for qualities, and while that opens opportunities for informed understanding, it can also create expectations or preformed judgments that may inhibit new perception.

With this caveat, antecedent knowledge is seen as a starting point, a step into the research environment that contributes to the ability to observe a situation, assess qualities and significance, construct new understanding, and then bring that perception to life for others. "Generally, educational critics examine a school's curriculum ideology (beliefs about what schools should teach, and for what ends) by focusing on the school's major dimensions (intentions, curriculum, pedagogy, school structure and evaluation). Today educational connoisseurship and criticism is used by educators worldwide for both research and evaluation" (Uhrmacher, 2001, pp. 248–249).

Typically, Eisner sees educational criticism as a function of the dimensions of *description, interpretation, evaluation*, and *thematics*, four elements that are neither independent of each other nor sequential in nature. Although it is necessary to discuss each of these dimensions one at a time, Eisner does not intend that they be viewed as imposing a rigid order or progression. They "do not prescribe a sequence . . . further, [they] do not imply that each is wholly independent of the others" (1998, p. 88). They offer tools for the researcher to use, not rules that must be followed, and each has been adapted in some measure in the gateway model.

Educational critics rely on a discerning eye for observation and a literary style to communicate what is seen in a manner that allows readers to feel they were present at the time. This descriptive account is characterized by an evocative clarity that enables the reader to know what the scene looks like and what it would feel like to be there, in essence, re-creating the event instead of just writing about it. "Detachment and distance are no virtues when one wants to improve complex social organizations or so delicate a performance as teaching. It is important to know the scene" (Eisner, 1998, p. 2). To empower a sort of vicarious participation, the researcher relies on the deep knowledge of qualities and their significance and then uses descriptive imagery, sensory language, and artistry to communicate the nuances of that experience to those who have not lived it. The purpose of this type of description is, quite simply, to help others understand.

Description of the pervasive qualities and characteristics observed in a setting or circumstance helps establish the connection between the reader of an educational criticism and the context in which it is played out. This function, however, cannot be achieved without foundations of interpretative thinking, for it requires not only an adept portrayal of what is seen and heard but also the discernment of what is relevant and informative. As Eisner explains, "Although there is no sharp and clear line to be drawn between the descriptive and interpretive, there is a difference in emphasis and focus" (2002, p. 229). While description provides an *account of* an event or situation, interpretation *accounts for* it (Eisner, 1998). It is not the details and facts of the event itself that are important but the interpretation of those facts that illuminates significance.

Interpretation, which asks what the situation means to those involved and what concepts can help explain it, requires a contextual basis for comprehending what was observed along with knowledge of antecedent factors that reveal possible causes, outcomes, and impacts. Through interpretation, theories can guide perception and offer a foundational tool to be used in explicating the possible meaning of what is seen. This key point is that there is not just one way to interpret and understand what is seen or heard, but many. "What theory provides is not single-minded, certain conclusions regarding the meanings secured . . . but rather frameworks that one can use to gain alternative explanation for those events" (Eisner, 2002, p. 230). This antecedent knowledge combined with solid grounding in a variety of theories and concepts prepares the critic to account for the situation and foresee potential outcomes or consequences, a key requirement for a gateway study.

For evaluation, the educational critic uses his or her ability to appraise qualities and discern which experiences have fostered growth, which have inhibited growth, and which have had no effect one way or another (Uhrmacher & Matthews, 2005). It is this aspect of educational criticism that addresses the essential *value* factor of educational practice. Education, as a normative societal function, ideally contributes to personal development as well as collective well-being. Evaluation helps improve educational practice by disclosing those factors or conditions that actually help or hinder the educational process (Eisner, 2002). Making judgments regarding the contributory quality of what is seen or heard requires that criteria for judgment be employed. Evaluative thinking from diverse perspectives on the complexities of the educational experience contributes to deeper discussion relative to the benefits and potential drawbacks of policy, practice, programs, and systems. However, Eisner does not contend that there is only one set of criteria or one perspective for what is of worth:

> The fact that two critics might disagree on the value they assign to a common set of educational events is not necessarily a liability in educational evaluation; it could be a strength. For much too long, educational events have been assessed as though there were only one set of values to be assigned to such events. (Eisner, 2002, p. 233)

For educational criticism, the disclosure of *thematics* serves as the "distillation of the major ideas or conclusions that are to be derived from the material that preceded it" (Eisner, 2002, p. 233). Looking for the prevalent lesson(s) that can be learned, thematics reveals the patterns or qualities from the situation being researched that can connect with other settings or situations. Themes or patterns that are observed in a situation are considered to determine what can be learned that extends to other settings, thereby providing a lesson that may be useful to others. "The point of learning a lesson is that it is intended to influence our understanding or behavior; it has some instrumental utility" (Eisner, 1998, p. 104). Thematics not only captures the essence of an educational criticism but also makes it possible to use that criticism in understanding other educational situations.

In-depth inquiry into experience and the meaning that people take from it is a generative process of appreciation and learning. "Human knowledge is a constructed form of experience and therefore a reflection of mind as well as nature. Knowledge is made, not simply discovered"

(Eisner, 1998, p. 7). In the strategies of educational criticism, Eisner provides an inspired method for discovery and meaning-making to researchers in education and in related disciplines. From its roots as an arts-based perspective for improving instructional practice, educational criticism has contributed to a variety of studies for a variety of purposes. Research as diverse as studies of Waldorf education in the United States (Uhrmacher, 1991), family visits to a Canadian science center (Munroe, 1997), literature and reflective practice in a Swedish secondary school (Elmfeldt, 1997), the use of technology in English language classes at a private school in Mexico (Morgan, 2007), flexibility grouping in an Austrian *hauptschule* (Carlile, 1993), the schooling of Navajo children in New Mexico (Holyan, 1993), and environmental education for the global society (Moroye, 2007) are but a few examples of its widespread applicability. These studies explore complexities of educational experience, presenting not only what happens in schools but also what it means to those involved and what contributes to the growth that is sought.

"Educational critics and critics of the arts share a common aim: to help others see and understand. To achieve this aim, one must be able to use language to reveal what, paradoxically, words can never say. This means that voice must be heard in the text" (Eisner, 1998, p. 3). Closely connected to Eisner's stance is oral history, which elevates the significance of voice to a prominence as in no other field of study, for as a methodology, oral history thrives through voice and expression. This is also the connection to gateway research, which uses oral history interviewing strategies to collect data that can be studied through the lens of the educational critic.

Oral history

The sharing of stories is an ancient custom in cultures worldwide, but it wasn't until the mid-twentieth century that asking individuals about their life experiences was recognized as a research methodology. Writers chronicling the development of oral history often remind readers that the first oral historian was Thucydides, who in the fifth century B.C. sought out eyewitnesses to interview and then used their stories to write the history of the Peloponnesian War (Yow, 2005). The collection of memories and personal testimony has, of course, continued across the ages, but the invention of functional recording devices in the twentieth century greatly increased the ability to capture and collect interviews so that they could be heard by others not present at the time. This development made the use of recorded testimony for research possible; however, the widespread use of tape recorders became practical only after World War II, when portable recording devices became available (Yow, 2005).

In 1948, Allan Nevins launched the first ever oral history project at Columbia University. Nevins was the first to employ a systematic effort of recording and preserving recollections that were deemed historically significant and then making those recollections available for future research purposes (Shopes, 2002). From that day until this, advances in technology have supported and advanced the recording, transcribing, archiving, and sharing of the human side of history.

More recently, the emergence of feminist research, postmodernism, postpositivism, and similar research perspectives has focused heightened attention on the complexities of experience and the meaning-making process, turning from a scientific model that quests for facts to support a reified truth and committing to an exploration of the varieties of experience and diversities of perception. With attention to the effect of power relationships on the research environment, qualitative researchers have shown a growing interest in narrative investigation, and many have turned to the techniques provided through oral history to investigate diverse perspectives and understandings. Voices from the elevated, from the mainstream, and from the marginalized are being joined to present a more complete representation of the broad range of human experience. By collecting the voices of individuals with living memory of events and situations, oral history helps bring awareness of personal biography as it interfaces with societal processes. With its power to democratize the voices that are heard, oral history enables researchers "to get at the valuable knowledge and rich life experience of marginalized persons and groups that would otherwise remain untapped, and, specifically, offers a way of accessing subjugated voices" (Hesse-Biber & Leavy, 2006, p. 151).

The term *oral history* has been used to describe a variety of undertakings. It, of course, involves the criterion of oral-ness, namely that a speaker tells something to a listener who, by prompting the "telling" and asking questions of interest, contributes to shaping the story of experience. Assumed within the term is that the interview is recorded for use either by the interviewer or by others. The term also carries a sense of historical significance, without restricting *history* to matters only of national or global importance by embracing the stories of lived experience of a more personal, individual nature. Oral histories include everything from formal interviews with heads of state and leaders of multinational corporations, to informal, facilitated conversations with aging ancestors for family lore, to community initiatives designed to capture a sense of place and shared identity with voices of the past, to admittedly fictionalized accounts based upon real stories told by real people. It is also used to describe in-depth interviewing conducted for a variety of research purposes in a variety of settings.

The Oral History Association (2000) promotes oral history as a "method of gathering and preserving historical information through recorded interviews with participants in past events and ways of life" (p. 4). In many cases, the recordings and accompanying transcripts are not published but instead are archived in libraries or online repositories. "They are like silent memories waiting for someone to rummage through them and bring their testimony to life" (Fontana & Frey, 2005, p. 709). These interviews, whether published or archived, provide a means to access memories of events and situations as a function of personal interpretation of experience. In turn, they enable analysis of individual and collective lenses on the effects of situations on people and the ways that circumstances are understood by those who were there. "The way in which people make sense of their lives is valuable historical and societal evidence in itself (Oral History Society, 2007, para. 5).

An oral history interview asks individuals to talk about their life experiences, to tell their own story without being subjected to interrogation requiring confirmable details of cognitive recall or demanding absolute content accuracy. "Oral historians consider hearsay, opinion, beliefs, value judgments, and even errors as part of the peculiar usefulness of oral sources" (Portelli, 1991, p. 256). Indeed, the purpose of oral history is to ask individuals who have shared some experience, location, or moment in time to tell their stories and the meanings they take from them. It is assumed that what they choose to tell is what they consider most important and relevant to the questions that are asked. If the narrator doesn't introduce matters that are of interest to the interviewer, a question may be asked that will prompt consideration of that area and elicit the commentary that is sought. However, if the narrator chooses not to respond or answers in a cursory manner and moves to another topic, that behavior in itself constitutes data.

A hallmark of oral history is its reliance on memory "not only as a source of details but also as a rich repository of thoughts, beliefs, and impressions of self-understandings and historical understandings that have evolved over time" (Clark, 2002, p. 572). Marked by the relationship between the story-teller and the story-hearer, it explores how individuals interpret their circumstances and the way that their understandings change (Portelli, 1991). What is shared in an interview

depends on that narrator's memory, interest, and personal understanding as well as his or her literal and metaphorical vantage point
relative to the event.

Oral history allows learning about people's lives from their own point
of view, whether they were directly involved in a situation, were on-scene
observers of the situation, were connected to the individuals directly
involved, or were associated more distantly. The experience as described
from each vantage point will differ, as will the interpretations of the
event and its perceived consequences. Similarly, understandings about
the situation will differ in relation to the individual's prior experience
and awareness. A person with a previous connection or involvement will
experience a specific event differently from one who has never had exposure related to the event being studied. For example, a labor union
organizer will experience a transit strike differently from an urban com-
muter; a veteran school teacher will experience school reform differently
from a policy maker; and a homeless person will experience urban
renewal differently from a city planner.

When oral history interviewing is used for investigations with guiding research questions, interviews must be analyzed so that findings
can be drawn. The purpose of the investigation influences how the
researcher will analyze the data, but analysis always involves an extensive and deep review of the transcripts as well as the audiotapes and
any ancillary reports, documents, and artifacts.

Oral history's holistic approach deals with a story as a context-
based event, not a detached set of happenings or a source of discrete
data to be coded as abstractions. Just as conducting the interview
requires interaction and rapport with the narrator, so too the process
of analyzing life stories calls for informed interaction and connection
with the data. In spite of the time-consuming demands of this ven-
ture, analysis is not a task that can be performed by a computer-based
software that attempts to quantify qualitative information. Removing
the contextual basis in order to standardize responses and apply
organizing codes for computer analysis runs the risk of decontextual-
izing the meaning and complicating the construction of generative
research findings. "Because meaning is contextually grounded . . .
coding depends on the competence of the coders as ordinary language users. Their task is to determine the 'meaning' of an isolated
response to an isolated question, that is, to code a response that has
been stripped of its natural social context" (Mishler, 1991, p. 3).
Coding depends on an assumed relationship between language and
meaning. Since human experience differs, especially with regard to the
unique functions of language, regionalism, cultural background, and

event-specific knowings, a reliance on this assumed relationship for computer-based analysis is shaky at best.

Given this state of affairs, how then does one analyze oral history interviews? The answer to this question is that many approaches to analysis are available. In fact, the same interview could be analyzed in a variety of ways, since the approach is dependent on the disposition and particular interest of the researcher as well as on research purpose that guides the investigation. Yow (2005) cites an informative set of examples for oral history analysis, suggesting that life history can be analyzed to discern the differing roles that individuals play in events, their personal characteristics and modes of adaptation, developmental history, inner life, significant turning points, psychological matters, patterns and their relationship to theory, historical documentation around a topic or event, symbolic thinking and representations, and so on. Whatever lens is used, the effort will require close attention to the content of the interview, the language that is used (this includes silences and nonverbals as well), the disclosures that are given and those that are withheld, and the alignment with or divergence from other accounts and records. As you can see, the task of the oral history researcher is much more than to record enjoyable conversations with interesting people.

The point I would like to repeat here is that oral history as a form of narrative inquiry is all about voice, the voice of the narrator and the voice that the researcher uses to present the narrator's story to the listener or, in the case of written text, the reader. Even though oral history researchers tend to minimize their own voice and advance the voice of the narrator in communicating the life stories, it is the researcher who shapes the story and then faithfully shares it with others.

Poetic display

Concern for the voice of the narrator as a means to preserve that individual's reality and experience has prompted researchers to develop alternative means of reporting interview data. In the early 1990s, Laurel Richardson began experimenting with unconventional forms of data presentation, finding prose unsatisfactory for the representation of life stories in her sociological research. Seeking a means to bring artistry and readability to academic writing, Richardson (1992) considered the responsibilities that attend the writing of people's lives:

> When we write social science, we use our authority and privilege to talk about the people we study. No matter how we stage the text, we—the authors—are doing the staging. As we speak about the people we study,

we also speak for them. As we inscribe their lives, we bestow meaning and promulgate values. I concluded that the ethically principled solution to issues of authority/authorship/appropriation required using my skills and resources in the services of others less beneficially situated. My conclusions were satisfying as rhetorical, aesthetic, and philosophical abstractions; but how to write substantive sociology that pleased me was still elusive. (p. 131)

Richardson's solution to this dilemma was to use the interviewee's words in the construction of poetry. "Louisa May," the life story of an unwed mother, is a 3-page poem that Richardson crafted from a 36-page interview transcript, using only her narrator's words, expressions, images, speech patterns, and rhythms (1993). This work demonstrates the great potential for the genre to connect and create an aesthetic as well as emotional response, bringing a narrator's experience to life on the page. Richardson, who references the poem as both "data" and "findings," states, "In writing sociological findings as poetry, I felt I had discovered a method which displayed the deep, unchallenged constructedness of sociological truth claims, and a method for opening the discipline to other speakers and ways of speaking" (1993, p. 697). Louisa May's interview data as a poem models a way of telling that creates in its readers and listeners an emotional response in addition to a cognitive knowing.

Richardson also considered what part *writing* plays in qualitative research, since "language is a constitutive force, creating a particular view of reality. This is as true of writing as of speaking, and as true of science as of poetry," adding, "*How* we are expected to write affects *what* we can write about" (Richardson, 1991, p. 174, emphasis in original). Acknowledging that she, like most of us, had been taught to start writing only after knowing what she was going to say and in what order, she felt constrained by rules that reflected "mechanistic scientism and quantitative research . . . a sociohistorical invention of our 19th-century foreparents" (Richardson, 2005, p. 960). Seeing writing itself as an inherently creative process that required both interpretive as well as expressive processes led her to conclude that *writing* should not be seen as the end product of an investigation, but as a "method of inquiry," a "method of knowing," and a "process of discovery," offering a "practical and powerful method for analyzing social worlds" (Richardson, 1996).

Others, of course, had also seen the value in this means of representation. Anthropologist Dennis Tedlock, for example, used poetic forms to present his work among the Zuni Indians of New Mexico, feeling that poetry is closer to speech than is prose. In a paper presented to the annual meeting of the Organization of American Historians in 1973, he wrote:

> Nobody, whether in a
> literate society or not **speaks** in **prose**
> unless he is
> unless perhaps he is
> reading aloud
> from **written** prose
> and in the flattest possible voice.
> The **worst** thing about written prose is that there is no **silence** in it.
> (Tedlock, 1991, p. 113, emphasis in original)

Alessandro Portelli, whose search for folk songs in Terni, Italy, led him to a career as an oral historian, echoes a similar concern about the inability of a transcript to capture spoken communication: "A thorny problem in oral history is the translation of speech (especially non-standard speech) into writing. Transcription cannot avoid obliterating some meaningful marks of regional, class, or personal identity and history" (Portelli, 1991, p. 83).

In Richardson's attempt to capture a meaning that prose transcripts or summaries could not convey, she took great advantage of the power of poetic form to convey emphasis, rhythms, and silences. Her transformation of interviews into poetry served as inspiration for Corrine Glesne, who adapted the approach into a slightly different product, a device she terms *poetic transcription*:

> In poetic transcription, the researcher fashions poem-like pieces from the words of the interviewees. The writer aspires to get at the essence of what's said, the emotions expressed, and the rhythm of speaking. The process involves word reduction while illuminating the wholeness and interconnectedness of thought. Through shaping the presentation of the words of an interviewee, the researcher creates a third voice that is neither the interviewee's nor the researcher's but is a combination of both. This third voice disintegrates any appearance of separation between observer and observed. (Glesne, 1999, p. 183)

Glesne's interest in alternative representation of research occurred at about the time that she was interviewing Dona Juana, an 86-year-old professor of education in Puerto Rico, and coincided with her growing interest in writing poetry. When she read through the transcriptions of Dona Juana's interviews, she was inspired to experiment with poetry, describing a process that is similar to traditional data analysis: "After reading and re-reading interview transcripts, I generated major themes, then coded and sorted the text by those themes. My desire to create varied 'portraits' of Dona Juana helped guide the development of themes" (Glesne, 1997, p. 205). She then used Dona Juana's words to convey the essence of the interview transcripts as well as their emotional impact and manner of expression.

Whereas Richardson's poetic representation aspires to *be* poetry, "poetic transcription is similar to poetry in its form and use of concentrated language but it may or may not arrive at the artistic sensibilities of a good poem" (Glesne, 1999, p. 187). Its value is not as an achievement of the poetic in a literary sense, but in its ability to focus attention on the key and commanding aspects of interview data. Poetic transcriptions allow the reader to step into the shoes of the other and make connections, analyze the familiar in fresh ways, and reflect on circumstances in ways that allow ambiguity, difference, and openness (Sparkes, Nilges, Swan, & Dowling, 2003).

Rather than continuing to tell you about poetic transcription, let me show you by providing an example from my own work. The following examples are taken from an interview that I conducted with a student ("Sophia") who had discontinued her doctoral studies but was returning to the university to complete her dissertation, the only remaining requirement for her degree. The first excerpt (Exhibit 4.1) is taken directly from Sophia's interview transcript. Exhibit 4.2 is a prose summary of the same section of the transcript. Exhibit 4.3 is taken from the 3½-page narrative reduced in the style of poetic transcription from her 18-page interview transcript.

The excerpted transcription does more than tell the high-points of the content of Sophia's interview; it gives access to the significance behind the simple facts. Showing her former love for writing, her sense of loss for a pastime she once enjoyed, her discomfort in writing for academic purposes, and her feeling of having her creativity rejected in favor of rules and regulations, the transcription brings to life an aspect of the experience of this young woman, one of many students who remain at the ABD (all-but-dissertation) level.

Reducing interview data by excerpting and sequencing a narrator's words and phrases brings the reader close to the data and compels notice, focusing attention so that the essence and meaning of the interview become alive and accessible. However, not all interviews lend themselves to being displayed poetically. Even when the content of the interview makes this seem a promising strategy, the purpose of the research may dictate another form of presentation altogether. As Glesne notes, poetic transcriptions are not always appropriate: "It depends on the inclination of the presenter, the nature of the data, the intended purpose for writing up one's research, and the intended audience" (1997, p. 218).

Just as there are many methods for research, there are many ways to report data. Poetic form is a viable method, but for some researchers, the term *poetic* may be sufficiently intimidating to discourage them from even making the attempt. Be assured, however, you do not need to be a poet, or even poetic, in order to use the gateway approach.

Exhibit 4.1   Excerpt from Sophia's interview transcript

CM: *Do you like to write?*

Sophia: Yes, I really did like to write when I was a kid—it was one of the things that I thought I was going to go on to do. I wrote little books when I was a kid and I remember enjoying that and reading—well I had a hard time when I was a kid with reading but once I really learned how to read—and I liked to write. . . . I don't write anymore.

CM: *You don't?*

Sophia: No.

CM: *Was that by choice or by chance?*

Sophia: I guess I got—a couple of things stopped me from writing—I'm not good at critical writing. But, I remember a book I wrote for school and I got tremendous praise for it. It was this little story, and I'm sure I don't have it any more, but I do remember the story. It was about a little adopted girl and I wrote this whole story and did the illustration and I kept thinking, you know I want to be a writer. . . . I want to write—I loved it. And I think I also liked the feeling of giving back so in my head I always thought I was going to be doing that. I think along the way I found out that I'm not such a great writer. It was cute at that time, but I'm not so sure I was reinforced thereafter. . . .

CM: *But it was something you enjoyed back then . . .*

Sophia: [Hesitates] I feel I wanted to—yes.

CM: *It's not something you're afraid of? You're not afraid of writing?*

Sophia: I've always been taught—not taught—I always had the feeling that I was discouraged because of—not because of the creativity but the grammar—the importance of all of the rules and regulations versus the creativity. . . .

CM: *So that's sort of . . .*

Sophia: That stifled and then put me at a point that—I think I became somewhat sensitive—I'm not good at criticism and I take it personally—a writer needs to be able to learn from that, to learn from feedback. It sets me back. . . .

I could write—but there's different ways of writing—like academically—I think academically as I went into the doctoral program I think trying to—I think purposely—take my fun at writing and put it into the academic world and it didn't work—they weren't accepting of it. I was at one point going to write a whole doctoral dissertation—one of my papers was through the eyes of one of my deaf students and following that. They really didn't want any part of that, so the creativity in me along with the creativity in that world was not one that was accepted. . . .

Exhibit 4.2 Summary of Sophia's interview data

As a child, Sophia had enjoyed writing and had been quite proud of a little book she had once written in school. She had seen writing as a way of giving back, but her creativity was stifled by grammar rules and regulations, and she now feels insecure about her ability. When she entered the doctoral program, she tried to bring her enjoyment in writing into her academic work, but her work wasn't accepted. She now feels insecure about her ability to write for academic purposes.

Exhibit 4.3 Poetic transcription

I wrote little books when I was a kid
I remember enjoying that and reading—
I don't write anymore.

I wrote a book for school
A little story, about a little adopted girl
I wrote this whole story
I kept thinking
I want to write—I loved it.
Liked the feeling of giving back.
Along the way I found out I'm not such a great writer.
It was cute at that time,
but I'm not sure I was reinforced thereafter.

I was discouraged
Grammar—rules and regulations versus creativity.
I'm somewhat sensitive—
Not good at criticism
It sets me back

I went into the doctoral program
Wanted to take my fun at writing
put it into the academic world.
It didn't work—
They weren't accepting of it.
They really didn't want any part of that,
the creativity in me

Member check

Lincoln and Guba (1985) describe the process of *member checking* as the most critical technique for establishing credibility of research because this is the step in which data, analytic categories, interpretations, and conclusions are tested with study participants and stakeholders. Ranging from daily check-ins with informants to confirm accuracy of the data being collected, to more formal interactions where respondents are asked to review the analysis and conclusions, this process is a vital component of sound research practice. "If the investigator is to be able to purport that his or her reconstructions are recognizable to audience members as adequate representations of their own (and multiple) realities, it is essential that they be given the opportunity to react to them" (Lincoln & Guba, 1985, p. 314).

Member checking is both a formal and informal practice as well as an ongoing means to check the dependability of the data and the researcher's understanding thereof. In one regard, member checking happens on a daily basis. It can be achieved by simply asking study participants to correct contradictory or inaccurate information, to provide additional or illustrative examples that clarify the account, or to help the investigator come to understand the situation more clearly, from that person's perspective.

Lincoln and Guba suggest several structured ways to conduct a member check: "A summary of an interview can be 'played back' to the person who provided it for reaction; the output of one interview can be 'played' for another respondent who can be asked to comment; insights gleaned from one group can be tested with another" (1985, p. 314). For a comprehensive member check, Lincoln and Guba advocate scheduling an in-depth session, perhaps lasting one or more days, to give representatives of all stakeholder groups the opportunity to review the investigator's presentation of facts as well as interpretive accuracy, and to respond, agree, disagree, or provide additional confirming or contradictory information. However, the confirmation that is sought through member checking is not to be confused with triangulation:

> Triangulation is a process carried out with respect to *data*—a datum or item of information derived from one source (or by one method or one investigator) should be checked against other sources (or by other methods or investigators). Member checking is a process carried out with respect to *constructions*. Of course, constructions may be found to be noncredible because they are based on erroneous data, but the careful investigator will have precluded that possibility by virtue of assiduous earlier triangulation. Member checking is directed at a judgment of the overall credibility, while triangulation is directed at a judgment of the accuracy of specific data items. (Lincoln & Guba, 1985, pp. 315–316, emphasis in original)

In interview research, member checking offers narrators a chance to become a part of the interpretive process (Glesne, 1999). As a matter of form, most in-depth interviewers give their narrators a copy of their transcripts to ensure that the audiotapes have been correctly transformed into print. This provides narrators the chance to verify the accuracy of the record as well as to check for information that they would prefer not to have reported, such as identifying details or statements that might become problematic. It also gives an opportunity to extend conversation, and as a result "both the researcher and the researched may grow in their interpretations of the phenomena around them" (Glesne, 1999, p. 152).

Seidman (2006) suggests that in some cases member checking could cause difficulties for the researcher and advises caution against giving the participants too much control at this stage. Some interviewers may grant a right of review that almost equates to a veto overriding the researcher's ability to process, analyze, and write up the results of the investigation. Seidman notes that there is actually a continuum of interviewer–interviewee relationships, with those who consider the work to be a joint creation (implying co-ownership) on one end, and on the other, those who suggest a more autonomous relationship that ends with the interview, thus leaving the only incumbent responsibility as informing the participant about the nature of the research and then not distorting what is reported in the interview. The philosophical perspective on "ownership" will impact the degree of member checking that is deemed appropriate, and Seidman concludes that while he shares with participants any material that concerns them, he retains the right to write his final report as he sees fit, while taking into account issues of accuracy and vulnerability, of course.

While a traditional member check provides a means for confirming credibility, it is not necessary that a researcher honor all concerns and disagreements expressed by participants or stakeholders or to rewrite or reinterpret what is in the record to accommodate distractions such as personal agendas, biases, or uninformed vantage points. Nor is it incumbent on the researcher to add, delete, or abridge the account in an attempt to homogenize the data to get at the average of what participants have expressed. What is required, instead, is that the researcher attend to the feedback gained through the member check and to assess its relevance and its worth. In this way, the member check contributes perspectives additional to what was gained through the data collection stage.

One more point before leaving the issue of member checking: I have heard graduate students and even their instructors advise against taking the step too seriously because it poses the potential risk of having data "pulled" at the last minute by study participants.

My thought on this issue is that the purpose of research is to acquire, document, and analyze the participant's understanding. Without confirmation of the accuracy and completeness of the data, the research is of questionable value. Yes, it is possible that participants will disagree with your rendering or even pull their data from the study. That is their legally protected right. However, if you treat them with respect, give them the opportunity to see how their contributions to the study are being understood, and allow them to clarify confusing or contradictory information, I think they are less likely to withdraw. Knowing about and correcting inaccuracies or misunderstandings prior to publishing an investigation can avert the damaging consequences of such disclosure at a later date.

PUTTING IT TOGETHER

Having considered the structures that support the gateway, it is now time to consider how the approach works. While aspects of the approach have been inspired by other traditions and strategies, they are re-envisioned and reworked to create something new. Elements of epistemology and purpose in educational criticism and oral history harmonize in this approach, and the adaptations of poetic transcription and member check add dimensions that contribute other ways of knowing and disclosing. Table 4.1 summarizes key traits of the two qualitative methodologies along with the characteristics of the gateway approach so you can distinguish similarities and differences.

As shown in Table 4.1, oral history records events, experiences, and circumstances from the perspective of the people who have lived them. Serving the purpose of historical documentation as well as narrative study, oral history interviews capture narration in a speaker's own voice. Resultant audiotapes may subsequently be indexed, transcribed, or archived in the original form. The purpose is to reveal meaning and significance that individuals take from their experience. While educational criticism also focuses on meaning and disclosure from the participants involved in a particular setting or activity, its primary purpose is to study a set of school-related circumstances or practices for the purpose of improving education. In this regard, educational criticism has an action component inherent within it. Oral history seeks to document and understand, but not necessarily to stimulate action. Gateway research seeks to deepen the understanding of circumstances and situations with particular attention to the meaning that people take from their experience; it emphasizes the value of keeping the participant's voice present in the research so that future actions and decisions can be better informed.

Oral history methods can be used to acquire information about any lived experience, so too can both educational criticism and gateway, though the educational critic focuses on experience within an educational context. Oral history primarily examines past events and experiences; educational criticism looks to observable events taking place in the present; and gateway could be used for research on both past situations as well as ongoing (present) circumstances.

A depth of knowledge (a connoisseur's knowing) is especially important to educational critics and gateway researchers because of the need for refined discernment of qualities. Oral historians and gateway researchers rely on interviews first, with secondary consideration of resource documents and artifacts. Educational critics tend to rely on observations, with interviews and consideration of documents or artifacts as secondary. Whereas all three pay attention to subtleties of perception, oral history and gateway listen to nuances of sound and silence, while educational critics attend more to the visual, making meaning from what is observed.

For member check, an oral historian asks interviewees to read their transcript and confirm its accuracy. An educational critic confirms accuracy of the research through structural corroboration with other sources, consensus validation with other researchers, and referential confirmation by those who read the work. A gateway researcher asks narrators to confirm accuracy of the transcript, to assess the excerpted narrative for its accurate portrayal of the experience and interpretation of meaning, and to reflect on any added understandings that may have come through the research process. The term *narrator check* is used to describe this function.

Oral historians commonly index or transcribe interviews and produce a written record that may then be analyzed for historical or sociological research purposes; they also may publish in a variety of academic and literary journals. Educational critics describe, interpret, evaluate, and thematicize the collected data. They may write up their study for publication, but they also attempt to speak directly with the educators and other significant stakeholders who participated in their study so that they may learn directly from the research. Gateway research produces individual and collective narratives that make the data accessible and demonstrate the basis for research findings, but tapes are not necessarily archived.

For the purposes of gateway research, open-ended interviews focused on lived experience, characteristic of oral history, provide a means for

Table 4.1 Characteristics of oral history, educational criticism, and the gateway approach

	Oral history	Educational criticism	Gateway approach
Purpose	To document events, experiences, and the meaning people take from them. Less a quest to establish absolute fact than to uncover meaning and significance	To describe, interpret, and evaluate educational settings, teacher practice, or other events for the purpose of improving education	To deepen understanding of the complexities of human experience and meaning for the purpose of informing decisions or shaping action
Research topic	Lived experience over a period of time	Educator practice within context of culture and processes	Impacts/consequences of experience, situations, policies, decisions, etc.
Perspective	Narrator's reflection on past experiences, events, and circumstances	Observation of events or situations occurring in the present	Narrator's reflection on past or ongoing experiences
Data collection	Data generated collaboratively. Primary method: open-ended interviews. Additional data collected from primary and secondary source material and artifacts. Attention to nuances of sound	Researcher as instrument. Primary method: observation. Additional data collected from follow-up interviews, artifacts, and primary and secondary source material. Attention to nuances of sight and image	Researcher as instrument and collaborator. Primary method: open-ended interviews. Primary and secondary source research to build appreciation for insider perspective. Attention to nuances of sound and sight
Data confirmation	Interviewee confirms accuracy of audiotape and transcript	Confirmation through other sources (structural corroboration), other researchers (consensual validation), and readers (referential adequacy)	Narrator check to ensure correct interpretation of data presented in excerpted narrative

Data display	Archived audiotapes, transcription, or audio presentation	Artistic verbal rendering evoking visual imagery and sensory perception	Excerpted narratives from transcripts of interviews
Data processing and analysis	Transcription of taped interviews to produce a written record that may be analyzed. Life history analysis, linguistic study, narrative analysis. Understandings about experience and meaning	Description, account of setting and events. Interpretation that accounts for what was seen and contextual meaning. Evaluation of what fostered growth, inhibited it, or had no effect. Thematics to create story extending beyond the situation being studied	Interpretation from a strong contextual basis, discerning factors and consequences of events and experience to create excerpted narratives. Evaluation of factors that fostered growth, inhibited it, or had no effect. Analysis and reporting of patterns/themes across all participants
Presentation and dissemination	Archival of tapes and transcripts for future research (access may be limited to researchers). Publications in a variety of fields (e.g., history, anthropology, sociology)	Raw data rarely archived. Publication of summary data in qualitative research and writings for educators and policy makers	Archival of tapes possible but not essential. Publication of excerpted narratives and findings in a variety of fields and for a variety of purposes

discovering information that a narrator considers significant enough to remember and recount. Data display incorporates the dimensions of description and interpretation from educational criticism; namely, it takes the form of evocative narratives that are created by interpreting and distilling the participants' transcripts. Interpretation is supported by antecedent knowledge and relevant research that contributes to the conceptual or theoretical framework. This process actually occurs concurrent to the creation and presentation of the narrators' stories that serve as a gateway to understanding the experience. When consistent with the research purpose for a gateway study, evaluation can be built into the research focus by expressly framing guiding questions that address the narrator's perspective of positive and inhibiting factors related to the experience. Through analysis, the researcher identifies patterns and themes across all narrators and calls attention to the importance of the events and what meanings those events may have in other settings.

THE GATEWAY PROCESS

While particularly useful for researching topics that may be difficult to study (e.g., discrimination, victimization, exploitation), the gateway approach also contributes a means to learn from situations of a more routine nature (e.g., school reform, policy implementation, community development). This type of research seeks to deepen understanding of human experience and to facilitate discernment of the meaning and significance that people take from their own lives. This increased understanding serves a utilitarian purpose, potentially making things better by prompting action to improve a situation and, at a minimum, by contributing to a shared appreciation for the complex experiences of humanity.

The completion of a gateway study involves six general processes: preparation, interviewing, interpretive display, narrator check, analysis, and reporting. Guidelines for completing these are given in the following chapters, but it is important to remember that these processes are not independent of one another, nor are they neatly sequenced in a lock-step fashion. It is quite likely that the research will proceed in a spiraling pattern with some overlap and flux as the data emerge and the findings begin to take shape. In general, however, these processes may be described as follows:

- *Preparation* requires clarifying the purpose of the research and the primary research questions and building requisite knowledge and appreciation of the qualities of the experience so that it is possible to research from an informed perspective.

- *Interviewing* involves a series of three modified oral history interviews to disclose meaning and significance that the narrator takes from the experience being investigated.
- *Interpretive display* of the narrator's story requires the discernment of salient information to answer the research question(s) so that an excerpted narrative can be created to disclose the data in an accessible and evocative manner.
- *Narrator check* achieves a consensus that the excerpted narrative accurately presents the narrator's data and intended meaning and that both the researcher and the narrator have come to a shared understanding of the experience and its significance for the narrator.
- *Data analysis* entails the search for patterns, themes, and understandings across all narrators, utilizing techniques adapted from educational criticism.
- *Reporting* is that essential, final step in which results of the inquiry, namely the conclusions that answer the research questions, are shared in ways that achieve the intended purpose for the research while preserving each narrator's voice.

With foundational knowledge of a situation being investigated, the gateway researcher enters a community of experience to collect data through purposeful and informed in-depth interviews with those who have lived that situation. Acquiring a true insider's understanding prior to the research might be impossible to achieve. However, with dedicated effort it is possible for investigators to deepen their knowledge of the situation and be able to ask discerning questions and listen for meaning in a way that might be difficult without appreciation for the subtleties of the experience and the potential responses to it.

Depending on the purpose of the research, it is common for the gateway researcher to interview multiple individuals so that multiple narratives and therefore multiple perspectives can be shared. Gateway is not about finding the average of what has been experienced, but instead seeks to illuminate the diverse expressions and responses to that situation. Like educational criticism, "it is a matter of being able to handle several ways of seeing as a series of differing views rather than reducing all views to a single correct one" (Eisner, 1998, p. 49).

After transcribing the interviews, the researcher reads, rereads, considers, interprets, and distills the interview data from a strong contextual and interpretive basis, discerning relevant factors, consequences of circumstances, and feelings resulting from the experience. As the "qualities" of the experience are considered, themes and patterns emerge. Research questions that guide the investigation make possible the reduction of data and the creation of excerpted narratives (via strategies from poetic transcription) to present data in a way that evokes what it felt like to be a part of that experience. Since narrators are asked to share the positive as well as the negative aspects of their situation, it is possible to identify common themes and factors. In research directed at learning from experience, the fundamental narrative quality of the display engenders stories that offer access into the experiences of the participants and that invite others to learn from this world of meaning and significance.

Before data analysis, it is essential that all narrators review their individual narrative and meet with the researcher to discuss their responses to it. This checkpoint is the safety switch assuring that it is the narrator's voice and not the researcher's voice that is heard. In distilling the narrative and sequencing or arranging phrases and expressions, the researcher runs the risk of subjectively shaping the data because in making choices of what to include and what to leave out, meaning can be changed. By going to the narrator and saying—*This is what I understood from your interviews; did I capture what you wanted to communicate?*—a researcher circles back to confirm and authenticate the piece.

The need for this step partially stems from the dynamics of normal conversation since people do not organize their thoughts chronologically or thematically. The human brain simply doesn't work that way. Instead, we communicate with dangling thoughts and fragmented memories that may be completed or embellished later in the conversation. The researcher attempts to make those associations and connections of meaning, and for the delicate task of sequencing an excerpted narrative, it is important to make sure that thoughts and stories are connected correctly. A narrator check thus enhances the mutuality of understanding by confirming the accuracy of the contents and assuring that both the researcher and the narrator reflect on the meanings as accurately interpreted. In research terminology, the narrator check validates the data.

In an investigation involving multiple narrators, this process of confirming accuracy of all excerpted narratives is essential before a "cross-case" analysis can be completed. For analysis, the task is to look for repeated elements across all narrators' stories, those patterns that shed light on the questions guiding the research. For example, through analysis, you can draw credible conclusions about positive factors that narrators have pointed to as beneficial in a situation or those that are seen as having a negative effect. You will likely see differences across the range of your narrators, with some narrators feeling one way while others feel quite differently. Indeed, that is why you interview more than one narrator; you want the opportunity to consider different perspectives and to learn more about what contributed to these differences. When you draw conclusions, you will be able to report your findings from a basis of contextual understanding that communicates the "inside story" of the experience as viewed by different individuals.

In completing a gateway study, you will find that you do not move through these six processes in a linear progression, since it is quite likely that you will find yourself working on more than one part of the project at a time, perhaps interpreting one narrative while confirming accuracy of another, or beginning the analysis only to discover that you may need to reconnect with one of your narrators to clarify a point that you had not noted previously. Since gateway research is built from practices considered throughout this text, I have included in Appendix B-2 a checklist of steps for the approach, along with a reference to the chapter(s) where each is discussed. It is my hope that this guide will help you plan and complete gateway research in ways that will achieve a harmony of purpose, a soundness of design, and even, in the scientific use of the term, a certain elegance of implementation.

Criteria for gateway research

The criteria for in-depth interview research are of course relevant to a gateway investigation. Sound gateway studies evidence validity through accuracy of meaning confirmed by each narrator, replicability through transparency of implementation, reasonableness of results, disclosure of researcher relationship to topic as well as subjective responses during the study, utility of the findings/conclusions, and achievement of the purposes they set out to serve. In addition, an assessment of the quality of a gateway study would address characteristics more specific to the approach itself:

• Research is conducted from an informed perspective with antecedent knowledge illuminating all parts of the process; specifically this includes the study design, selection of narrators, data collection, data reduction (narratives), interpretation, narrator check, analysis, and reporting.
• Analysis is supported by a conceptual framework to facilitate discernment and explication of patterns and their significance.
• The researcher self-reflects regarding responses to the topic and narrators.
• The researcher has conducted a meaningful narrator check with each participant, achieving consensus that the excerpted narrative

faithfully represents what was expressed in the interview as well as the intended meaning.

- The narrator check has provided the researcher and the narrator an opportunity for mutual understanding.
- The excerpted narrative is internally consistent and, if not, inconsistencies are reconciled.
- Vividness of excerpted narrative to present the data provides access to experience that underlies the findings and conclusions that are drawn.
- Findings are supported by the data.
- The research contributes to the understanding by others of a situation or circumstance that they might otherwise not know.
- Research has utility in deepening understanding so that practice and decision making are better informed.

By employing the voice of the study participants who contribute their individual accounts, gateway research evokes an understanding of events and perspectives on a deeply personal level, making accessible the human impacts and consequences of circumstances and events. Through a focused narrator check, the gateway researcher confirms that the narrator's story has been correctly heard and interpreted, thereby contributing to the validity and replicability of the study. While not the purpose of the research, gateway seems to produce an added benefit for narrators in deepening their awareness of their own understandings about the experience and personal responses to it. Analysis considers the meaning of what has been learned from the study participants, and through reporting, the significance is communicated to an audience to expand and deepen their understanding of that community of experience so that the basis for the conclusions is clear.

Implications for the researcher

The following four questions are frequently asked. To avoid any misconceptions about this type of research, I will answer them directly before moving on:

1. *Do I have to be a poet?*
   No. The gateway approach does not require skills in writing poetry or even a love of poetry, just an appreciation for human experience and fluency in communication. An excerpted narrative isn't poetry—it is a data display that distills and sequences particularly meaningful words and phrases from an interview transcript in order to make it accessible to readers and to transform it into a coherent presentation

that can be analyzed for research. White space around the phrases does make it look like poetry, but its intent is to focus and direct attention, highlighting the essential elements among the many words used to describe an experience.

2. *What topics are appropriate for gateway research?*
Gateway can contribute to historical or documentary research as well as to educational and social science research designed to inform and prompt action. Investigations that seek to explore human experience, social change, public policy, personal challenge, or program effectiveness are among the many that could be explored through this approach. Since gateway addresses the significance of qualities, research to identify quantifiable factors such as graduation rates or cost/benefit ratios would not be appropriate topics.

3. *Do I have to write a complete life story for every interviewee?*
No. While in-depth interviewing practices common to oral history are used to collect the story of the narrator's experiences related to a certain situation or circumstance, it is not necessary to conduct a complete life review. Oral history traditionally builds a contextual background for the narrator's life experience, perhaps beginning with a question about place of birth or childhood memories. Gateway targets an event or circumstance and then pursues relevant life history that relates and informs understanding of what emerges in the narrator's telling of the event. An oral history can be a huge continent; a gateway study is a small country, and an excerpted narrative does not need to be a complete biography.

4. *How do I know if I would be able to "do" a gateway study?*
Rather than answer this question for you, I'll ask you to decide for yourself. Your research question and purpose should drive your decision to employ any specific research method, but a self-assessment of your disposition and personal comfort with the requirements of research must be considered as you make this decision. Consider the questions in Table 4.2 on page 76 to see if this approach is a possible match for your research endeavors.

What's Next?

Learning from the experiences of others can be a thoroughly human and humane endeavor. As a researcher, your task is to gently ask important questions and confirm your understanding, and then analyze and

Table 4.2 Aligning researcher interest and research approach

If you want to research . . .	for the purpose of any of the following . . .	and you want to learn from . . .	and you're comfortable with . . .
• meaning and significance of human experience • events and experience from an "inside" perspective • human experience in a natural setting (i.e., not a laboratory)	• deepening understanding about a circumstance, event, or situation • communicating in-depth story and analysis • informing policy, decision making, program design or implementation • documenting life experience on many levels—mental, physical, emotional, psychological, behavioral, spiritual • humanizing the research (putting a face on the numbers)	• fewer data sources providing more in-depth data • personal narratives • generative expressions and interpretations of experience • authentic voice(s) of those involved • varied perspectives and understandings	• an emerging design • ambiguity • openness, lack of structure • data overload • resource demands (time, energy, cost) • attention to both the obvious and the elusive • self-reflection • being your own devil's advocate • building rapport and interacting with diverse individuals • returning to narrator to confirm accuracy of your understanding • writing and analyzing meaning through language

then the gateway approach may be a viable model for your research.

report your findings in ways that acknowledge the narrator as the source of information and insight.

With a sense of the purpose of gateway inquiry and an appreciation for the research traditions that contribute to its practice, you are ready to begin your investigation. Chapter 5 focuses on the preparation stage, upon which all else is built. In this stage, you clarify your purpose and your research questions, acquire background knowledge, develop a framework to support the inquiry, and prepare for interviewing.

Chapter 5

Preparing for the Research

Chapter Topics:

- conceptualizing the research
- building your capacity as a researcher
- the proposal process
- getting ready to interview

Research of any kind requires a great deal of preparation. As you begin your investigation, you need to ready yourself for a pursuit that will take months or, in some cases, years to complete. Gateway inquiry is no different from other methods in this regard, for good research requires planning, contemplation, and building a capacity to achieve the research purpose.

Preliminary Steps

Earlier chapters introduced you to some of the decisions you will make as you begin your study. Briefly, to review them here, a prerequisite step to using any approach is to first assess why you want to research and what you want to learn. Illuminating the purpose of the research involves clarifying how you envision your work being used and forces a consideration of possible misuses that should be guarded against. From the very start, you must become your own, most critical devil's advocate. By forcing yourself to consider alternative interpretations of your work, you are more likely to focus your efforts and avoid misunderstanding.

As you work through the *why* and the *why not* of your research, you will also refine the primary research question to guide your investigation.

Research questions that are a good match for a gateway study are along the lines of the following:

- What is the experience of . . . ?
- What are the long-term effects of . . . ?
- What are the outcomes of participation in . . . ?
- What characteristics influence the development of . . . ?
- How do individuals adjust to the life event of . . . ?
- How did planners reach the decision to . . . ?
- How can a program be designed that meets the needs of . . . ?
- How do children respond to changes in . . . ?

Underlying broad questions such as these are more specific secondary or defining questions. For example, if you want to identify factors that contribute to a successful program implementation, you might want to frame sub-questions to specify that you are seeking information on challenges, supporting conditions, inhibiting factors, immediate versus long-term change, perspectives from varied vantage points, developmental stages, or such. By focusing on several sub-questions, you contribute layers of understanding to your broad overriding question.

Determine your research questions very carefully, especially if you are conducting dissertation research. While you may truly desire to answer a significant question on the large scale (e.g., *Why do students fail to complete high school?*), you really don't want to spend the rest of your life researching such a huge topic in order to complete your degree. You may indeed dedicate your professional career to solving some of society's most troubling problems, but you can do that after you've graduated. For the purpose of dissertation research, reduce your question to the smallest, definable elements. Be specific and don't get tripped up by attending to the grand-scale questions that really trouble you. Instead, start small; target a discernible, specific issue and population, and then, after you've earned your degree, take on your life's work. For example, the question about students' failure to graduate might be reduced to a manageable, *How does implementation of [a specific] district policy influence the decision of Hispanic boys to stay in school and graduate?* Sub-questions might focus on challenges, workable strategies, recommendations that can be shared with other students, and potential resources or changes that are needed. By breaking down a problem into smaller units of focus, you are, first, more likely to complete the study, and second, more likely to produce findings that point to solutions.

In deciding on a research topic, you really do need to focus on something that you're interested in. You will be living with the research for a long time, so consider the expertise and knowledge that you already have. What are you curious about? Where are your interests, your strengths in making fine-tuned discernments, your prior knowledge, the questions that you want answered? This type of self-reflection can be a starting point that leads you to a research topic aligned with your personal interests, foundational knowledge, disposition, and professional curiosity.

Often, as graduate students are searching for that elusive research question, they tend to overlook a manageable project close to home. Educational and social science research attempts to find solutions to problems faced in the field of practice. Look around at the situations you observe in your immediate world. What could you investigate that would contribute a clearer understanding and perhaps lead to the development of a new practice or improvement of an existing one?

If you are an independent researcher, not constrained by the demands of satisfying a dissertation committee, the task of clarifying a purpose and focusing research questions may come more easily. However, it is important not to minimize the significance of this step. Even though you know what you want to do and why, it is helpful to put it in words, on the page, as a guide and a reminder of your specific goals for the research.

By the time you arrive at a workable research question, you will probably have an idea of the general target group or population for your inquiry. Limiting that population in ways that get you to the point of your study can be a challenge. For example, you might have real interest in learning from the experience of gifted students, or homeless people, or veterans. You cannot possibly complete a study comprehensive enough to meet such broad ambitions. Instead, think small and specific: What segments of those general populations could you address? Find ways to limit the descriptors of the sample. For example, the above populations could be refined to girls with an IQ over 145, homeless women, or Vietnam veterans who received treatment for posttraumatic stress disorder (PTSD).

Reviewing studies by other researchers will help you further clarify your research population. After becoming more informed about the issues faced and the domain of knowledge to be explored, you will be able to refine your question and specify your sample. The study group considered in the previous examples might be reconceived as high school girls with an IQ of over 145 in rural school districts, or homeless mothers of school-age children in the inner city, or physically disabled Vietnam veterans who received treatment for PTSD within five years of their discharge from the military.

Building conceptual frameworks

For interview research, you need an idea of what is known about your topic before you can reasonably determine how you can contribute to that body of knowledge. Constructing meaning requires a solid understanding of the broad concepts or theories related to the experience (which will become your conceptual or theoretical framework) as well as an appreciation of the specific situation or experience being studied. To know the field, it is necessary to become familiar with the theoretical knowledge that applies to your study of the particular situation or event as well as the functional knowledge of those who have lived the experience.

Both theoretical frameworks (built from relevant theories that attempt to explain behavior, processing, and response) and conceptual frameworks (based on less developed concepts that symbolize related ideas) provide support for understanding what can be learned from the experiences of others. A framework specifies "the main things to be studied—the key factors, constructs or variables—and the presumed relationships among them. Frameworks can be rudimentary or elaborate, theory-driven or commonsensical, descriptive or causal" (Miles & Huberman, 1994, p. 18). In effect, frameworks define the boundaries of the territory to be explored and provide a guidebook pointing out what to look for and how to make sense of it. This is where background research—which you will synthesize as part of your review of the literature—comes in.

A framework of concepts or theories focuses attention and, as a result, "what we experience is shaped by that framework. Thus, the questions we ask, the categories we employ, the theories we use, all guide our inquiry. . . . Language shapes, focuses, and directs our attention; it transforms our experience in the process of making it public" (Eisner, 1998, p. 28). The framework helps in translating that language so that you and others can begin to analyze what it might mean. Building a framework and scaffolding research with a strong review of literature will focus your lens of inquiry. Framing education and social science research doesn't require finding every scholarly publication that has ever been written on a topic. Instead, it requires deconstructing the research question and its intent into its key elements, breaking down the question into parts so that it is possible to determine which theories, concepts, or background knowledge will be useful for interpretation and analysis. (Appendix A suggests a variety of reference and bibliographic tools to assist you in this stage of your research.)

There is no one right framework or body of literature for any given research question. It is a matter of the justifiable arenas of knowledge that a researcher chooses to employ in the shaping and/or supporting of the analysis and the drawing of conclusions for the study. For example, for the Columbine study, I built a framework from a synthesis of what was known about lethal school violence, the exposure to traumatic events, and family functioning. Moroye's (2007) investigation of ecologically minded teachers drew upon environmental and ecological education literature as well as Eisner's school ecology. A study of parental involvement of Mexican immigrant parents in the schooling of their children employed a framework constructed from critical race theory, Latina/o critical theory, a construct of race/ethnicity, and its negative impact on educational opportunity of worth with connections to democracy and social justice (Prosperi, 2007).

Of equal importance in preparing yourself for completing research is an appreciation for the details and qualities of the specific experience or situation under investigation. The gateway researcher operates like the educational critic, seeking to say "useful things about complex and subtle objects and events so that others less sophisticated, or sophisticated in different ways can see and understand what they did not see and understand before" (Eisner, 1998, p. 3). In order to say those useful things, it is essential to have the skilled appreciation and antecedent knowledge of the subject's "subtle and complex qualities" as well as "the conditions that give rise to these qualities," since knowledge of those conditions influences the ability to experience the qualities (Eisner, 1998, pp. 63–64). This background knowledge informs the investigation and helps the researcher to recognize and interpret the factors that are perceived. "Learning to see what we have learned not to notice remains one of the most critical and difficult tasks. . . . Everything else rests on it" (Eisner, 1998, p. 77).

To make the most of an interview, it is necessary to be informed enough about the topic and the setting so that questions can be well framed and appropriately posed. At the same time, it requires openness to discovery and inquisitiveness that come from sufficient background knowledge free from the presumption that all answers are already known.

Acquiring an inside perspective

To understand the terrain you will be exploring, you need not only to build your knowledge and understanding of the topic in general but also to develop your appreciation of culture in the broad sense as well as the culture specific to that experience, language and all of its uses, and a range of common and atypical responses. Doing this preparatory work enables you to ask meaningful interview questions, hear with deeper comprehension, and access perspectives that contribute to your being able to achieve the goals of your research. "Unless researchers . . . understand the cultural values of another, they may fail completely to understand the significance and meaning of the testimony. Turning on a tape recorder is not enough. Informed listening is an essential adjunct" (Harris, 1991, p. 6), and that requires listening from another's perspective.

Developing an insider's perspective involves being able to speak the same language, using words with the correct or accepted meaning within that community of experience. For example, what do you call the event, situation, or person? Do school administrators bristle on hearing gang violence being referred to as a *school* problem, or do they refer to it as a *community* problem? Is the term *handicapped* child offensive? Is the student who chooses not to graduate a *drop-out*? Do teachers in the school consider classroom volunteers a *distraction* or an *asset*? Simply knowing the language of the experience, not an academician's rhetoric, is a vital component of a successful interview. If you don't know the language specific to the research environment, I encourage you to connect with others who can help you build this fundamental component of communication before you start your research interviews.

In some cases, matching characteristics of race, gender, ethnicity, ideology, or social class of the population you wish to study can offer you a subtle advantage in being seen as *simpatico*, or sharing similarities and interests that will facilitate communication and understanding. However, an alignment of personal history or demographics is no guarantee that you will be able to record the narrator's lived experience and understand its implications with an insider's knowing. Likewise, being from within the community of experience or having lived a similar event or situation does not ensure that you will be able to see from *that* narrator's perspective. Each person views and responds to a situation in a uniquely individual way, and it is your task to record and study what your narrator knows and feels. Tacit assumptions and judgments need to be set aside.

While being accepted into a research site may be easier for a true insider, in many respects, research from within the setting becomes more challenging, for it requires overcoming your personal lens in order to understand from the other's point of view. In this situation, the practice of *reflexivity* can provide insights that will assist in differentiating self from the narrator while remaining aware of the connectedness that exists. "Reflexivity is the process through which a researcher recognizes, examines, and understands how his or her own social background and assumptions can intervene in the research process," by reflecting on differences as well as similarities to the individuals whose experience is being researched (Hesse-Biber & Leavy, 2006, p. 140). Your personal background will influence your perceptions, so awareness of your own experience and cultural lens can help disclose areas where you may unintentionally taint the data or simply not see it. Your job is to learn from the other's perspective and not cloud it with your own. When it comes time for the interview, you will need to adjust your level of "knowing" and allow your narrator to teach you. It is the narrator who holds the knowledge that you lack.

When you have learned more about the topic and find yourself feeling more and more connected to a particular area of investigation, consider beginning a research journal. Developing a greater familiarity with your research topic will, not surprisingly, be the start of a remarkable journey of discovery. When you write your research proposal and later when you report your findings, you will need to disclose your relationship to the topic, your background, and your subjective connections to the research questions and study participants. This disclosure of who you are as an investigator is standard practice in research, since a reader needs to know how to judge your work and evaluate your findings. Readers need to know, for example, if you are conducting an inquiry into a program that you designed so they can assess subjective relationships that might influence your research. This does not mean that you cannot conduct an inquiry that is close to home, only that you must assure your reader of the measures you have taken to monitor for bias.

By starting a journal at the beginning of your trek, you will be able to mark your own progress, observe your reactions to the research, record the challenges and successes you meet, and analyze your own experience related to the study. Consider the credentials that you think prepare you for this work, contemplate the areas where you feel unprepared, reflect on your hopes and fears as you begin the process, document unsettling as well as validating experiences and interchanges, and monitor significant responses to what you are learning. When it comes time to describe the "researcher's relationship to the topic" for your research proposal, you will be well prepared, for you have tracked your development all along the way.

A research advisor, either a dissertation chair or a trusted colleague, can help you refine your thinking about your investigation, pointing out difficulties that you may face, noting strengths of your design, offering suggestions for resources and support, and extending your own thinking by providing a different perspective. Consider setting aside a section in your research journal to record these thoughts.

DEVELOPING AN INTERVIEW GUIDE

To make the most of your interviews, it is important to consider how you anticipate your narrators contributing to your research, what you want them to discuss, and how you want the interviews to proceed. With your background knowledge you are able to determine the areas that you want to explore in the interchange and then use that information to draft trial questions, focusing on broad inquiries to invite the narrators to tell their story. You may know a great deal about the situation, but you do not know what your narrators know and how they feel about it.

In order to ensure that you address the critical topics with each of your narrators, you need to design a simple interview guide to remind yourself of the matters that you want your narrators to speak about, with several open-ended questions that you can use to initiate conversation along the lines you want to pursue. The exact flow of the questions or topics will differ for each of your narrators, and in fact, you might find that you won't need to ask some of the questions at all; your narrator may provide that information in connection to a different question. After developing some tentative interview questions or requests, ask a colleague to help you by doing some role-playing and interviewing you using those questions. You can discover how specific wording influences or directs your response and use this experience to help you polish your interview techniques.

I think it is helpful to include your research question on your guide to remind you of your focus and to spell out exactly what you will tell each participant about the research, so that you establish uniformity in presentation. Just to be sure there's no confusion, you are not going to ask the narrator to answer your research question—that's what you're setting out to discover. However, as you ask open-ended questions to invite narrators to share their stories and insights, it is helpful to keep your focus on the intent of your interview, which is to collect information that you will use to answer the research question. Also remember that an interview guide is for *your* use; it is not something that you hand to your narrators and ask them to respond to. That's more characteristic of a questionnaire or survey.

An example of this open-ended approach to questioning is provided in Exhibit 5.1. For this hypothetical research, let's assume that an investigator has identified a sample of former students who participated in an experiential literacy program for second language learners at imaginary East Shore High School. The investigator is guided by the research question, and for the first interview, has decided to consider the following general and open-ended prompts.

This type of questioning provides the essential ingredients for an interview guide to define the territory you want to cover in your complete series of in-depth interviews. Such a guide helps organize your thoughts so that you address the issues you want to explore. Its purpose is not to spell out every question that you will ask, or restrict the scope of your narrator's responses, or constrain you from discovering new terrain that you might not have anticipated. An example of a guide for a series of three complete interviews for the East Shore investigation is provided in Appendix B-3 to demonstrate how this first interview forms the basis for subsequent sessions.

Exhibit 5.1 Sample questions

Primary research question: How does participation in an experiential literacy program affect the students who are enrolled?

Sub-questions: What are the benefits and what are the impediments to participation? What are the outcomes for students? What recommendations would students make for improving the program?

Questions for first interview:

- Ask, *I'm interested in learning about the experiential literacy program and would like to hear of your experiences in the program at East Shore.*
- From the resulting discussion, look for effects and ask follow-up questions concerning academic growth, socialization, family, self-esteem, career decisions, or other topics related to the research question.
- Ask about key points from the narrative, for example, *You mentioned that you didn't get along with some of the other students in the program and even felt threatened. Tell me a little about that.*
- Ask, *I'm interested in what you've been doing since you left high school. Can you expand on that?* (This is connected to outcome.)
- At the end of the interview, explain that you will explore some of these areas more deeply at the next session. Ask your narrator to make a note of anything that comes up in the time between the interviews that might be of interest.

As you plan your interview guide, which you need to include in your proposal to conduct the research, you might take a moment and develop data management tools, including, for example, an interview summary to help you manage and keep track of the information you need from each participant (see Appendix B-4). After an interview, you can use such tools to document the areas that have been addressed so that you can monitor where you are in the process and target areas that remain to be covered.

PROPOSING THE RESEARCH

Before you can begin your research and make contact with potential narrators, you need to submit a formal proposal to your research committee to cover the design and significance of your research project and, if that is approved, a separate proposal to your institution's IRB explaining your plans for participant protection (see Chapter 3). To complete these proposals you need to consider the viability of your research questions, the practicality of completing the research, its significance to the audience you have targeted, and so forth. These are the kinds of questions that antecedent knowledge can help you answer. With background understanding of the circumstance or community of experience, you will be better prepared to anticipate what you will encounter in the research environment. Also, your awareness of the situation will help prepare you to design safeguards to protect your study participants and yourself as you undertake the research.

When the time comes to submit your formal proposal to conduct the research, you need to satisfy your advisor and research committee that your design is solid, that the investigation is feasible, and that it poses no harm to your participants. Different institutions and organizations require different formats for the research proposal, and it is up to you to prepare a plan that meets the criteria that apply. However, the following details are among those most commonly required in a proposal to conduct research:

- An introduction to the problem, issues, or topic to be studied
- Your purpose for conducting the research
- Your specific research questions (primary and secondary)
- An explanation of the significance of the study and how you can contribute to the body of knowledge in your field
- A review of the literature to provide an overview of current understanding and to support your rationale for the research

- Theoretical framework or conceptual lens(es) that can be employed in analysis
- The methodology and design of your research, including the number in your sample, how you will locate and select your narrators, how many interviews you plan and their length, and how you plan to analyze your data
- Proposed interview guide
- Strategies for protecting study participants from harm and for safeguarding their confidentiality
- Your relationship to the study topic and any delimiting factors
- A timetable for completing your research.

At the same time you are crafting your proposal to conduct the research, you need to be planning your application to the IRB or human subject review panel, if you are conducting research through an institution or organization that receives federal funding. While your research advisors will assess the merits of your proposal with regard to criteria for quality research, the members of the IRB will evaluate your application on criteria related to your plan for ethical treatment of your study participants, how you will gain their informed consent to participate, and the measures you intend to set in place to protect their confidentiality as well as to protect them from physical or psychological harm. After you have received the approval of your research committee and have made required changes to your proposal, you can submit your application to the IRB to conduct human subject research. With that approval, you will, at last, be ready to begin your investigation.

FINDING AND SELECTING NARRATORS

In your research proposal, you identified a certain population for your study, specifying how you intend to select the individual(s) from that broad population to interview. Once the requisite proposals are approved, you turn to the task of recruiting your actual narrators. Wise and informed selection is critical to the success of your inquiry, and many texts treat various approaches to making these choices (e.g., Miles and Huberman, 1994; Yow, 2005; Seidman, 2006; Creswell, 2007). The important thing to remember is that for in-depth interviewing, you are not seeking a random sample but one that serves a purpose related to your work. You are also not seeking a large sample (perhaps only six to nine narrators) because your goal is depth, not breadth.

Since you cannot interview everyone who has knowledge or perspective to share about your topic, you need to evaluate your options very carefully to determine exactly whom you will interview and how you will connect with them. Success in this type of sampling "lies in selecting information-rich cases for study in depth. Information-rich cases are those from which one can learn a great deal about issues of central importance to the purpose of the research" (Patton, 1990, p. 169).

As you build your knowledge of the situation or the community you wish to study, you can identify key institutions and individuals who have an immediate connection to the topic at hand. From that basis, you should be able to locate someone within the situation who can serve as a "guide and translator of cultural mores and, at times, of jargon or language" (Fontana & Frey, 2005, p. 707). Such a guide can be a valuable resource for you, not only in accessing potential narrators, but also in helping you understand some of the dynamics of the setting, the barriers that you will face, and the particular conditions that you may encounter.

If you are researching a topic that is related to a school or other institution, you will probably need to complete an application to their equivalent of an IRB before you are granted permission to conduct your research. Once you have formal permission, you can approach a responsible party at the organization to help you gain access to those you want to interview. This individual (or in some cases a committee) may serve as a gatekeeper, controlling access to the setting. When you meet with a gatekeeper, you'll explain your expectations for the study, your purpose, your questions, the intended dissemination, and potential outcomes from your research. A gatekeeper can become your ally, vouching for your legitimacy within the setting and introducing you to people you need to meet. Take along a one- or two-page project overview to help you describe your proposed study and to help you explain the kind of interviewees that you are hoping to recruit.

In some cases, the gatekeeper will connect you to others within the organization who can best recommend the actual participants for your study. Ask these general informants to suggest people to interview relative to your research question. For example, after being approved to conduct research in a municipal Department of Social Services, you might ask an administrator of the division for foster care to identify social workers who will then be asked to recommend parents to participate in your study of the experience of foster-care providers. Or, after gaining clearance through the research office of a school district, you might be referred to a building principal who can help you identify teachers who have successfully negotiated the shift to block scheduling.

In working with an informant, you need to assess how credible or reliable that person's perspective is within the circle of investigation. For example, if you want to research the effectiveness of a district-mandated literacy program for second language learners, you should question whether the person who helped develop the program is the best one to recommend participants for your study. A desire to see the program painted in a favorable light might influence his or her recommendations. A more likely informant might be a grade-level coordinator, department chair, or principal who could suggest teachers or paraprofessionals to interview.

Follow all leads and be sure to document who made the referrals, so that if asked you can let your interviewees know how you got their name. For matters of privacy, your informant or gatekeeper might not want to give you the telephone number or contact information for potential interviewees and may want to check with them first and then ask those who are interested to contact you. In schools and social service agencies, this is commonplace, but don't rely on an informant to explain the details of your study to potential participants. You are the one who best knows about your plans for the research, and as a result, you are the one who should describe its scope and purpose.

To achieve your research purpose, you need to interview narrators who have lived the experience you are investigating. Selecting narrators requires that you know what is "story-worthy" in the research environment (Chase, 2005). It also requires that you find individuals who have a story to tell and are open to telling it to you. Given a choice between a participant who is marginally connected to a situation you are investigating, or one who has firsthand experience, you would want to engage the latter. Given a choice of two individuals with equal connection to the experience or situation, with one who communicates freely while the other seems uninterested or pressed for time, you might find interviews with the former will produce deeper discussion and more useful data.

Remember, you are not seeking a random sample. Randomness is a concept aligned with quantitative studies that involve large numbers of participants who are selected at random to represent an entire population. For in-depth interview research, your goal is to interview a small number of people who know a great deal about the topic you are investigating; thus you are developing a "purposeful" sample (Patton, 1990), a strategic one, a sample that is "information rich" related to your research questions. "Interviewees are not statistically representative of the population at large . . . they are selected, not because they represent some abstract statistical norm, but because they typify . . . processes"

(Grele, 1991, p. 131). Identify those who are knowledgeable and can articulate their experience, and rely on your own judgment, based upon what you have learned in the preparation stage.

While not seeking randomness, you do need to consider the range of experience that you intend to address. Do you want to investigate *typical* cases, in which you can learn from those whose experience was most common, or do you want to learn from *extreme* cases? Do you want a *convenience* sample, one that is easily accessed but may lack a depth of information and credibility? Do you want to begin interviewing and then ask narrators to recommend others for your study, relying on a *snowball* effect to help you complete your sample?

Consider how diversity or variation among participants would influence your research. Including members of different ethnic groups or socioeconomic classes enhances your investigation by bringing in a wider range of perception and response. Diversity may make the challenge of data analysis more time consuming, but you can deepen the understanding of an issue by providing access to a variety of views about it. "Reality is complex; to accurately portray that complexity, you need to gather contradictory or overlapping perceptions and nuanced understandings that different individuals hold" (Rubin & Rubin, 2005, p. 67). You will need to make choices like these from an informed vantage point and realize that you define the territory you want to investigate by setting boundaries and limits. Issues of time, resources, and access are significant factors, and each constraint that you accept impacts the scope of your study. Be guided by your purpose as you make these decisions.

As noted previously, in-depth interviewing is often characterized by an emerging design with a study sample that may evolve as your research progresses. In your first interview, for example, you may learn about other individuals who can provide additional or differing information. Your research questions as well as the parameters of your IRB agreements will influence your choices about whom to include in your study.

You will probably make the first contact with potential narrators by phone. Prepare detailed notes of what you want to say to ensure that you cover the key points in your first contact. In the preliminary phone conversation, introduce yourself, your credentials, and your purpose for calling in the broadest of terms. Don't attempt to explain your entire study over the phone, but provide enough information so that the potential narrator knows in general what you are asking. If the candidate is open to hearing more, set up an initial meeting to talk about your project in more detail and to assess how well that individual aligns with your research needs.

At the first meeting, you will explain who you are, what you're doing, and how you see them relating to your topic. Be sure to explain how their potential involvement will affect them, namely, how much time will be required, how the results of the study will be used and disseminated, how you will protect their confidentiality, and so forth. Be as thorough as possible as you describe their involvement. They won't sign the informed consent form until the first interview session, but you do need to give them a clear idea of what to expect. If they decline to be involved, be gracious in expressing your appreciation for their time and, if you are struggling to find study participants, ask if they perhaps know of someone who might help.

If you decide not to include a particular individual, a tactful explanation is in order. You don't need to go into the details, but a simple statement that you have a certain criteria or range of experience that you are researching and that you have met your goals will help them understand why you have not selected them for the study.

When you meet with gatekeepers, informants, and potential narrators for your study, remember that they may not share your dedication to the research effort. In fact, you are asking for blocks of time that might cause considerable inconvenience on their part. You are asking them to do you a favor. To help them see that this is a worthwhile effort, make clear not only the goals of your immediate research (e.g., to earn your Ph.D., or to complete a program evaluation), but also the broader value that their contribution might make. If you are researching, for example, because you want to make a difference or contribute to improving services for others, let your potential narrators know about your hopes for the study. In this way, they will be able to share in your sense of making a difference. Just be careful not to overstate potential outcomes or promise that their participation will lead to immediate changes or improvements.

Keep careful records of all contacts you make while conducting your study. A summary log—with names, contact information, who referred them to you, and the outcome of each contact—can be a valuable organizational tool to manage the flow of your research. I have provided a sample log in Table 5.1. In this example, you can see how notations are helpful reminders of action steps and considerations for scheduling future contact. A record such as this, whether maintained on your computer, your portable data device, or a three-ring notebook, will help you organize useful details as you complete your research.

Table 5.1 Sample contact log for potential study participants

Summary contact log: Potential narrators

Research focus: Experiential literacy program for second language learners

Ref. #	Contact info.	Background info.	Recommended by	Date of contact	Comments
1	Myra Jones 100 Main St. Centerville, NW 00014 School phone: (555) 612-3456 Mjones@centerville.xyz	Principal East Shore High School	Dr. Wesley (advisor)	2/11/09	Will participate Will suggest interviewees ASAP **Follow-up by 2/18/09 if haven't received recommendations by then
2	George Chang Contact at school (555) 612-3456 gchang@centerville.xyz	Oversees experiential literacy at East Shore	Myra Jones	2/27/09	Attending conference this week, call back after 3/10/09
3	Joseph Maeda	Student—withdrew from program	Myra Jones		**Wait until principal makes contact before contacting
4	Tesa Harper Contact at school (555) 612-3456 Mharper@centerville.xyz	Teacher, exp. literacy program	Myra Jones	3/23/09	Will not participate— too busy getting students ready for state exams
5	David Martinez (555) 215-2904 Centerville Univ. Box 308 tamwil@centervilleuniv.xyz	East Shore graduate	George Chang	3/27/09	Will participate—liked program but had problems with other students

After you have selected your participants, continue to document and keep careful records. Basic organizational tools can assist you as you move through your study by keeping essential contact information readily available along with up-to-date notes on the latest interaction with each individual. I recommend setting up a separate file for each participant, such as the one in Table 5.2, with a code or reference number that you can use to label the audio-recordings and any related documents, including a copy of the informed consent. You need to restrict access to all records that contain identifying information

Table 5.2 Sample narrator contact form for use in documenting interaction with each narrator

Narrator contact and follow-up
Research focus: Experiential literacy programs for second language learners
Ref. # 2
George Chang
100 Main St; Centerville
gchang@centerville.xyz
(555) 612-3456

Date	Contact by	Outcome	Comments
3/21/09	e-mail	Wants to schedule meeting for early April, but needs to check calendar to be sure of date	Not sure he wants to do this
3/24/09	phone	Scheduled his first interview: 4/4/09, 5 PM; his classroom (204 W)	Seems more interested now that he understands what is expected **Be sure to take copy of district IRB approval—he won't talk to me unless sees documentation that I have district OK!
4/4/09	in-person	First interview completed, see interview tape/ transcript #2-1 Scheduled second interview for 4/11/ at 5PM	Eager to see programs improved His experience at East Shore will be useful **Recommended students from previous years in program— see journal notes, Session 1

related to your study participants. Storing contact forms and records in your private, secured files contributes to your ability to fulfill your commitment to protect confidentiality.

LOGISTICS AND DETAILS

In scheduling the interviews, make sure that your narrators agree to the length of the sessions you are requesting and the number of times you'd like to meet. If possible, at the first interview, set up a tentative time frame for the follow-up interviews, and then confirm each appointment as you progress. Plan on scheduling at least three interviews with each narrator, about a week apart, so that you'll have time to transcribe the session before returning for the next. This structure allows you to review the information that was covered, identify points that need to be clarified or expanded on, and determine a good starting place for the subsequent interview. In addition, since each interview builds on the one before, both in terms of content and rapport, it is important not to allow too much time to pass between your sessions.

While it is possible to complete an interview over the phone, for in-depth interviewing, you need to conduct face-to-face sessions that provide an opportunity to observe body language and nonverbal cues. In addition, in an age of multitasking, a face-to-face interview reduces the likelihood that your narrator will be answering e-mails or paying bills instead of attending to your interview questions.

Timing of interviews involves decisions regarding length and frequency for the interviews and also encompasses issues related to calendar time. Be aware of what else is going on in the world of your narrators that will shape the kind of interview you will be able to have. Consider, for example, that it is April and you want to interview Advanced Placement (AP) calculus teachers about their instructional practice. You may urgently want to finish your data collection but lose sight of the fact that AP instructors work with graduating seniors and April is one of their busiest months. As another example, assume that you have set up interviews with the head of the local health department about an initiative to immunize infants. The morning newspaper reports that the state has reduced funding to several health department programs. The quality of an interview could be dramatically impacted by this news, even if the program in question has escaped the budget reduction unscathed. The point is that you need to be aware of the implications of what is going on in the larger world and assess whether it is wise to proceed or to reschedule for a more promising time.

Another detail you need to consider relates to the setting for your interviews. As Yow (2005) points out, the setting can influence the content of the interview. Your interviewee might tell more classroom stories, for example, if you are sitting with him in his classroom after school and he takes visual memory cues from the environment. Likewise, if you are sitting in a narrator's living room, the experience that she shares might be more family-centered, more personally reflective of the home setting.

Interviews require an environment that is comfortable, quiet, and safe. Plan to meet at a mutually agreed upon location, making sure that it's a place where the narrator will feel free to talk and not be constrained by the surroundings. You'll need privacy and comfortable furniture, especially since in-depth interviews usually run for 90 minutes.

Whether the interviews take place in the participant's home, classroom, office, or some other agreeable locale, you want to be sure to avoid places where the background noise will make it difficult to get a high-quality recording. You also want to find a spot away from distractions and with minimal potential for being interrupted.

Be sure to document in your notes where the interview takes place. While on scene, jot down a few key images to stimulate your thinking about the general setting and environment. When you've completed the session and you're back at your home or office, write a more in-depth description of the setting in your research notes. Include sensory imagery—what did the room look like? (Was it a sunny room or were there deep shadows from heavy curtains?) What could you hear in the background? (Were there sounds of children on the playground or soft music down the hall?) What odors did you detect? (Were lilacs blooming underneath the window or could you smell food being prepared in the school cafeteria?) This information can be helpful to you as you set out to create participant profiles or introductions.

Remember that you want to contribute to the understanding of the experience you are investigating. Painting an accurate picture of the setting for the interview can help evoke understanding by showing readers the physical setting as well as the content of the narrator's words and expressions. Consider, for example, that you are interviewing a teacher in his classroom about his experiences with the implementation of a reform initiative. A spacious, sunny room, with modern desks, up-to-date technology equipment, and bulletin boards with cheerful images connecting to some aspect of the curriculum, conveys a message that is quite different from a room located at the end of an oppressively dark hallway, with dusty shelves of tattered books, dilapidated workstations, and a broken overhead projector lying forlornly in the corner.

WHAT TO TAKE TO THE INTERVIEW

At the risk of stating the obvious, I'd like to offer some suggestions about what to take to your interviews. It can be helpful to put all of your supplies in a "to go" bag that you keep at the ready so as not to overlook essential items when you head off to an interview.

For the first interview, you will need to take **two copies of the informed consent form** that has been approved for your research. Your narrator will sign one of these forms and return it to you, and the other is for your narrator's files since it provides an overview of the research, contact information for you and your advisor, a statement of recourse in case problems arise, and other details (see Chapter 3).

Your **interview guide** is, of course, essential, along with a **summary form** and any other **organizational tools** that you have developed to help you ensure that you get the information you need. Don't forget to take a **notebook** or pad of paper, along with several **pens or pencils**, so you can jot down observations and notes about the setting, nonverbals that won't be revealed in audio, key phrases that you might want to use in follow-up questions, summary comments, and any points that you need to follow up on before the next interview.

Since you will be audiotaping or perhaps videotaping the interview, you need to take your **recording devices**, along with **microphones**, and perhaps an **electrical cord** in case of emergency. I strongly advise you to take two recording devices, perhaps both digital or one analog and one digital, because the possibility always exists that a battery will fade or the microphone fail. By making a backup, you will gain a level of comfort in knowing that you won't lose part of your interview. In my own case, I had just completed a compelling interview with a narrator once, and as we were shutting down, she thought of something she wanted to add. I didn't realize it at the time, but when I restarted the cassette recorder, the tape jammed. If I hadn't also had the digital running, I might have missed one of the most powerful of her recollections. Since it might appear confusing, tell your narrator why you are using two recording devices. I can't imagine it being a problem; it shows your narrators that you value what they have to say and are being conscientious about documenting their reflections.

Be sure to use **fresh batteries** and take **a spare set for backup**, again, just in case. Using batteries is preferable to using a power cord, so that you won't be stymied by the lack of a convenient electrical outlet and you won't have to worry about interference from other devices plugged into the same line.

In addition to your recorder(s), you will want to take several **cassette-tapes** for any analog device, and take extra in case the interview runs a little longer than you had anticipated. I recommend using tapes of 60-minute length, since longer tapes are more likely to break. Label both sides of the tape for the interview in advance, with the date, narrator code, and whether it is Interview 1, 2, or 3.

Take a **small clock** or wear a wristwatch that is easy to read at a glance. You need to respect your narrator's schedule. In addition, as your interview progresses, be conscious of the passing of time so if you are using an analog device, you can turn the cassette over a few minutes in advance of it running out. It is greatly disruptive to the narrator who may have just started telling you of a significant event and then, since the lead ends of a tape don't record, having to repeat that in order to make sure that you've captured the story in its entirety. When you do stop to turn the tape over, check that the *record* button is pushed and that the tape is advancing. You can leave a digital device running as you take care of the cassette.

Use the highest quality of equipment that you can afford to buy or borrow, since your work depends on your being able to capture the interview in a form that you can work with. (See Appendix A-2 for suggestions on digital tools.) When you have acquired the recorders that you intend to use, become familiar with their operations and do a sound check before going off to your first interview. Check for sensitivity to sound and notice how placement affects the quality of the recording. Test your analog recorder to see how it handles its own vibration, and if necessary, take along a padded cloth to place under your recorder. One of the analog devices I once used in an interview dutifully recorded its own hum as it vibrated on a marble tabletop.

When you have assembled the tools that you will take to the interview, it is helpful to glance back through your research journal, reminding yourself of the purpose for your work and reconnecting with your research focus. It may seem unnecessary, since you, of course, know what you are setting out to do, but taking a moment to ground yourself can help prepare you to begin the next stage of your work.

WHAT'S NEXT?

Making the most of in-depth interviews requires an appreciation for a world that is bounded by the metaphorical membrane of experience. You have built foundational knowledge so that you can gain entry into that world, and, once there, attend to subtleties and nuances barely discernible while not losing sight of the broad landscape. Knowing the language and culture specific to the community of experience opens access to informative narrators who can guide you to deeper understanding. While the details your narrator recalls in an interview are but echoes of a larger world, your preparation for conducting in-depth interviews in this arena will help you make connections and see greater significance.

In Chapter 6, you will learn about the interview process. These practices are not limited to gateway research and can be used to great advantage in other interview-based methodologies as well.

Chapter 6

Conducting the Interview

Chapter Topics:

- the importance of relationship
- a series of three interviews
- asking and listening

In-depth interviews are conversations with a purpose, namely, to sit with another and learn what that particular individual can share about a topic, to discover and record what that person has experienced and what he or she thinks and feels about it. Interviewing for a gateway study involves a series of three, modified oral history interviews. Multiple interviews allow you to maximize the opportunity to build rapport and learn from the reflections of the informed individuals who agree to participate in your study.

Relationship and Rapport

With a lifetime of conversations preparing us, sitting in purposeful conversation for an in-depth interview would seem to be a simple task. However, the function of a social dialogue differs greatly from that of a research interview, and as a result, the dynamics are poles apart. In a social conversation, the interchange is usually more of a mutual sharing, likely characterized by balance and a back-and-forth pattern of communication. In a purposeful research interview, this is not the case, for it involves an exchange in which one person (namely you, the interviewer) seeks to be informed by and learn from another person (the narrator).

While many quantitative studies conceptualize the participants as research subjects, that term is somewhat uncomfortable to many qualitative researchers who view the *subject-object* dichotomy as distancing. In-depth interview researchers may see their study participants as *interviewees* or *narrators*, or perhaps *collaborators, respondents, informants*, or *reporters*, terms that convey the humanness of the individual whose experience is being considered. I personally prefer the term *narrator*, but you should decide what seems right to you.

This emphasis on the humanity of the participant brings the potential for a richness in perspective, but it also poses challenges in dealing with inequality and the range of differences that are encountered. Issues that arise from unequal status and power in society as well as differences in simple demographics command the attention of responsible researchers. There is reasonable concern for the ways that differences can affect interview relationships and, thus, can condition what is learned in an interview. A narrator with limited formal education, for example, may feel intimidated by a researcher who comes from an academic or institutional setting and as a result may be hesitant in offering opinions and candid responses. Women from patriarchal cultures may shape a different narrative when interviewed by a male researcher than they would if interviewed by a female researcher from their own culture. Narrators from a fundamentalist religion might answer questions differently if the researcher makes it known that she is of a different faith.

Perceptions about differences in status, power, background, and ideology can impact the quality and depth of an interview, and these perceptions, whether accurate or misinformed, constitute an integral part of the interview dynamic. Power differentials will exist, whether or not you intend it to be so. "Equality . . . cannot be wished into being. It does not depend on the researcher's goodwill but on social conditions" (Portelli, 1991, p. 31). Although you cannot make differences disappear, you can be aware of them and consider how your narrators might respond to you, from their point of view. Ironically, with regard to the content of the interview itself, it is the narrator who actually holds the greater power—he or she is the one who controls what is shared and what is kept silent. The narrator elects how to understand the questions and how to answer them. In this light, an interviewer needs to be aware of the perceptions of imbalance or asymmetry and assess what can be done to establish an equitable relationship that invites sharing.

As a researcher, coming to terms with your own role is an important step, for you are, in fact, a guest in the narrator's world, one who is asking for help in learning about a community of experience. Entering into an interview relationship with another is made easier if you operate in the role of an interested guest, one who seeks the opportunity to learn and appreciates the narrator's role as the host or guide who holds the experienced perspective that you need. In the role of a guest, you can achieve a spirit of naiveté and openness that allows you to "set aside your assumptions, pretensions in some cases, that you know what your respondents mean when they tell you something" (Glesne, 1999, p. 83). While you do need to demonstrate confidence that you know what you are doing, you also need to show a readiness to be taught by your narrator.

As a guest, you must come prepared and not expect the narrator to do all of the work. Being informed about the general "lay of the land" is part of the preparation for the journey, as is becoming aware of the social context, human dynamics, language, and general milieu of the situation or event. Background knowledge is an integral part of asking informed questions as well as being able to understand the answers that are given. But even if you enter the interview setting with an appropriate demeanor, that does not mean that the narrator wants or knows how to function as your guide or teacher. You will need to lay the groundwork that helps make this happen.

In getting ready for an interview, you have learned about the community of experience you are entering, including such basic factors as norms and appropriate behaviors. For example, you should know in advance how formal or informal you will be. If you are interviewing a school superintendent at the district office, you might choose to wear business attire, but when meeting with a lower income family at their home in a subsidized housing project, you would do better to dress more informally, perhaps in jeans and a plain shirt. You will also need to decide what you call each other: Will you address your narrator by title or by first name? How do *you* want to be addressed? Be sure to make these determinations based upon what is considered appropriate for the setting and the culture of your narrator.

Etiquette and customs differ from culture to culture as well as from situation to situation, and while you want to achieve rapport, be careful not to take an attempt to fit in too far. It may not be advisable to use the jargon or speech patterns of your narrators or to try to match their style of clothing. For example, if you are interviewing a student who has abandoned high school and is living on the streets with his fellow gang members, wearing the "colors" of the gang can lead to unfortunate results. If you are interviewing an elderly woman from the Deep South, trying to twist your New York accent into her melodious speech tones may be perceived as condescending. In some cases, it is more productive to acknowledge a clearly obvious difference as a way of inviting your narrators to teach you about their world. For example, you might say, *You know, I grew up on a farm and never attended school in the city. I am hoping that you might be able to share your experience so that I might better understand the challenges of urban education.*

While you want your narrator to feel comfortable with you and trust you enough to be candid in the interview, you may, on occasion, find it difficult to establish a connection. You simply may not like some individuals. There are differences in personalities, and the fact is, not everyone is likable. Some people are just plain difficult to be around. Some are brusque or angry; some are bigots; some have done some pretty awful things; and some just rub you the wrong way. You don't have to like them in order to learn from them. What you do need to do, however, is to turn to face, within yourself, what it is that you feel is prompting your reaction to them. Perhaps they remind you of someone you once tangled with, or maybe you just can't get past what you consider outrageous behavior or beliefs. Self-reflection and perhaps debriefing the session with an advisor may help you peel back the layers to your own reactions and enable you to suspend judgment so that you can complete a high-quality interview. It is important to attend to the narrator's perspectives and experience—that is why you are conducting the interview. However, if you find that even after deep soul-searching you are unable to proceed without prejudice, then I encourage you to act with integrity in deciding what to do.

Just as it is not possible to *like* every narrator, an equally difficult challenge develops when you admire them too much (see Yow, 1997). Remember that you are building a research relationship, not a friendship. Glesne (1999) points out three particular ways that the issue of friendship with the narrator can influence the process: (a) in selecting narrators because you want to work with those individuals to the exclusion of others; (b) in being denied access to one group of informants because of an affiliation with others; and (c) in the risk that you will censor your questions or that the participants will shape their answers to say what they think you want to hear.

If you choose to investigate from within your own community of experience, you need to be especially mindful in choosing your narrators and in conducting the interviews. Interviewing friends or colleagues can pose a challenge to the rigor of your work. Aside from the potential for bias and loss of perspective, the dynamics of the interview may be strained, for your narrator may anticipate a typical back-and-forth pattern of conversation, while you will be attempting something quite different. Any temptation to respond to their questions and enter into a dialogue will deflect the attention from the essential aims of your interview. You want to learn how they experienced something, not share your perceptions with them. If a narrator asks you a question, be polite and respond briefly, but don't talk too much about yourself. Continually monitor your behavior and remember that you are there as a researcher and not as a friend or colleague. You may need to gently remind your narrator about the purpose of the interview and that it in no way is a reflection of a prior or future relationship.

Regardless of whom you choose for your study sample, establishing trust and a respectful rapport is critical. From the beginning, concentrate on building the rapport that will help your narrators feel comfortable talking with you. When you get together for an interview, don't immediately pull out your recording equipment and start asking questions. Take a few moments to engage in social pleasantries before getting down to business. Remember, you are following this approach to your research because you are interested in the human side of an issue. You need to show that you are interested in what they have to say and not just wanting to get the interview over with.

A final comment on building rapport: Never promise more than you can deliver; never withhold essential information; and never imply that your relationship is more than it is. Chapter 3 considered the characteristics of principled research. These are fundamental to establishing a productive research relationship with your narrators.

THE THREE-PART INTERVIEW

The type of in-depth interview that I advocate for research into experience is built from the kind of questions used to collect oral histories. Cognitive interviews or analytic questioning serve other purposes, but to learn from past experience and situations, in-depth interviews into life events and their impacts provide the means to hear "what the participant has to say in her own words, in her voice, with her language and narrative. In this way participants can share what they know and have learned and can add a dimension to our understanding of the situation that questionnaire data does not reveal" (Lichtman, 2006, p. 119).

If you are using in-depth interviews for research purposes, you want to get beyond the simple facts that can be disclosed through a questionnaire or the details that are relayed in the frequently told story. This goal requires that you interview your narrator on more than one occasion. In the first telling of an experience, people often relate stories that they have told many times. This is their publicly expressed tale, perhaps memorized through frequent telling. At the first interview session, this may be all that you hear, and while it provides important context, it is unlikely to bring you to the deep understanding that you are seeking. In the first interview, you are building your awareness of that person's individual experience, adding to your foundational knowledge in general. If you stop with only one interview, you miss the opportunity to go deeper into the experience, exploring other levels of action, emotion, perception, and meaning. As Mishler (1991) points out, "The one-shot interview conducted by an interviewer without local knowledge of a respondent's life situation . . . in short, a meeting between strangers unfamiliar with each other's 'socially organized contexts' of meaning—does not provide the necessary contextual basis for adequate interpretation" (p. 24).

For in-depth phenomenological interviewing, Seidman (2006) advocates a three-session design: "People's behavior becomes meaningful and understandable when placed in the context of their lives and the lives of those around them. . . . Interviewers who propose to explore their topic by arranging a one-shot meeting with an 'interviewee' whom they have never met tread on thin contextual ice" (p. 16). Crediting the originators of this design, Seidman (2006) summarizes the basic structure and intent of the interview series:

> Dolbeare and Schuman (Schuman, 1982) designed the series of three interviews that characterizes this approach and allows the interviewer and participant to plumb the experience and to place it in context. The first interview establishes the context of the participants' experience. The second allows participants to reconstruct the details of their experience within the context in which it occurs. And the third encourages the participants to reflect on the meaning their experience holds for them. (p. 17)

A series of multiple interviews, scheduled fairly close to one another, allows the narrator to expand on aspects that may have been introduced in one session but abridged or slighted due to lack of time or because the narrator became distracted. For a gateway study, you are starting with a deeper antecedent knowledge, having built what Eisner considers a connoisseur's ability to discern qualities and nuances. Your plan for data collection would work within the structure for a traditional three-part interview but employ techniques from oral history and thinking from educational criticism; thus your interview series might look something like the following:

- *The first interview* is a chance to hear the narrator's description of the event or experience and to learn about the participant's specific background and relationship to the topic. Ask narrators for examples or stories, their feelings about or reactions to the experience and the changes it has brought. If you are looking for recommendations, consider telling narrators that at the final interview, you will ask what advice they would give themselves if they could go back to the beginning of the event or experience. Transcribe the interview before Session 2 to help inform your questions for the next session.

- *The second interview* allows you to invite deeper conversation regarding the specific experience being investigated, and by attending to seemingly casual comments, you can encourage narrators to expand their narrative, searching their memories for parts of the bigger story and life experience they may not have offered in the initial telling. Ask for more examples or stories to illustrate their comments. (Note: There may be occasions, especially when the topic being researched is of limited scope or complexity, when Interviews 2 sand 3 are combined.) Again, transcribe the interview and review before Session 3.

- *The third interview* provides another chance to consider different aspects of the story as well as to reflect on the experience, its effects, long-term meaning, and what narrators feel is essential to share with others. Ask narrators to describe what it was *like*, since metaphors communicate intensities and depth of meaning that simple narration of events may fail to capture. Also ask narrators what questions they expected you to ask. If you haven't already asked for that information, do so. At the end of the interview, remind the narrator that you cannot share all of the stories/experiences, but ask for guidance—which stories or points are the most important to include? Ask: *What would you be disappointed to see left out?* For a gateway study, a fourth interview is scheduled for a narrator check to confirm accuracy and completeness of the data display. (See "narrator check" in Chapter 7.)

In my research, at the end of each of the first two interviews, I suggest to participants what I hope we will be able to talk about in the next session, perhaps identifying particular comments or events that have been touched on in the interview. Each time, at the following interview, the narrators seem eager to share additional information or insights that they have recalled. They offer stories of related events and interactions that had not emerged in our prior conversation. At the first interview for the Columbine study, for example, I told the parents that at the third session I would ask them what advice they would have given themselves on April 20th that might have helped them get through the challenges of the ensuing years. As a result, at the second interview, parents wanted to tell me what they thought they would have benefitted knowing, and then at the third interview, each added to the advice that might have been helpful to hear, offering deepening insights and awareness each time.

While it might seem that only one interview would serve you well enough, two or three in-depth interviews give you a much better chance of gaining rich and informative data that otherwise would be missed. In the period between multiple interviews, the narrator will be thinking— most likely on a subconscious level—about what transpired in the previous interview and as a result may become aware of added details or illustrative stories that are relevant. This aspect of human nature and memory capitalizes on the mind's momentum in following through on a task. You can open doors for discovery by introducing topics that you would like to explore in subsequent sessions or by asking the narrator to consider and suggest related topics to bring to the next interview.

This process of subconsciously attending to a topic that is introduced but not fully explored relates to a psychological phenomenon revealed in 1927 by Bluma Zeigarnik's research into the effects of human needs and tensions. Zeigarnik gave participants in her study "a number of tasks and [allowed] them to complete some tasks while leaving others unfinished. She found that subjects remembered the unfinished tasks better, the 'Zeigarnik effect' " (Landrum, 1993, p. 92). The brain seems to remain connected to a task that is asked of it, even when the immediate circumstance no longer demands that it perform. I often think of this effect when, at 3 o'clock in the morning, I can suddenly remember the title of a song or the star of a movie or the name of my fourth-grade teacher's dog.

While a three-part interview design gives you the opportunity to explore experience and perceptions in depth, it is important to use the time wisely. If your narrator has told you about an experience and you invite deeper reflection, at some point, enough will be enough. Knowing when to turn your attention elsewhere can help eliminate frustration on your narrator's part as well as on yours.

Getting started

"A cardinal rule is to come to the interview thoroughly informed and then to let the subject do the talking" (Harris, 1991, p. 5). Your job is to ask the questions that will get the narrator talking along the lines that you want to explore. Your advance efforts to learn about the setting and to build your capacity to discern and distinguish qualities associated with the circumstance and the field of inquiry should prepare you to make the most of your time together.

After a few moments of small talk to ease into the situation, decide the best spot for your interview. As noted earlier, it helps if you have comfortable chairs and a table where you can set up your recorders, away from television, radio, or other noises. It is amazing the background sounds that recorders can pick up. It is best to place your recorders on a solid surface such as a desk or a table between you and your narrator. Position each device so that you can casually glance down and see that the recorder is functioning properly and that the *record* light is on, and, of course, place the device so that the microphone faces the narrator. It is preferable to use a separate microphone, but if that is not possible, you may find that the built-in microphone will work satisfactorily.

With the ubiquitous presence of cell phones and other electronic devices, it helps if you and your narrator agree to put these tools of modern communication on silent mode or turn them off altogether. If it is apparent that such a request would be inappropriate for your narrator, at a minimum, you should silence your phone as well as any alarms from your watch or personal data device. You don't want to interrupt the flow of the interview while you fumble for a phone or fiddle with your watch.

As you begin the session, do a quick sound-check to make certain that your recorders are working properly. When you are ready to start the interview, set your narrator at ease by reminding him or her of the purpose of the interview, a little of what to expect, and how long the interview will take. Answer any questions that arise. If a narrator wants to hear your story related to the topic you're researching, share it in the broadest of terms. This is reciprocity and it helps build trust, but you don't want to spend valuable interview time recounting the details of your own story.

Next, discuss the terms of the informed consent and ask the narrator to sign the required form as approved by your IRB. Explain to your narrators that they have the power to end the interview at any point and that the consent form will then become null and void. Remind them that you want to record the session and explain why and what will happen with the tapes. It is always a nice gesture to offer your narrators a copy of their tapes or transcripts for their own records and to share with others if they choose. If your interviews will produce records that will be archived, you need to ask your narrators to sign a *deed of gift* form that relinquishes copyright to you and grants you permission to deposit the tape and transcript into a designated archive. For information on archival practices and sample forms for your use, see Valerie Yow's (2005) *Recording Oral History*, Donald Ritchie's (1995) *Doing Oral*

History, The Oral History Association's *Evaluation Guidelines* (2000), or other reputable oral history resources (see Appendix A). If you are not planning to archive the records, you can tell your narrators that they may use real names in the tape and that you will edit out all identifying information for your transcript and data display.

Once you've determined that all is in order, start the tape with a simple statement of who you are, whom you are interviewing, when, where, and for what reason. For example, you might say, "This is Chris Austin, and I am talking with Jenny Bryan in her home on January 19, 2010, about her early years in foster care in the Blank County social service system. This interview is being conducted for dissertation research into how being in foster care influences long-term social development and life goals."

Asking questions

First interviews may begin a little awkwardly, with a degree of uncertainty and formality. As things progress, however, both you and your narrator will find increased confidence and comfort, and things will generally flow a little more easily. This is probably the first time your narrator has ever been formally interviewed. Remember how you felt in your trial interview when you asked a colleague to interview you? Even though you knew what to expect, you were probably a little hesitant and self-conscious.

Always start the first session for all participants in the same way, using the question that you decided on for your interview guide. This will ensure that you introduce the study uniformly, cover the preliminary details, and begin from the same point of departure. As your narrator answers the first question, that's when things diverge as you follow his or her answer to wherever it might lead.

The broad, open-ended questions that you outlined in your interview guide cover a familiar topic or content. Make sure you ask your questions in a manner that is clear but not condescending. Be open and inviting. As you move through the interview, remember that your interview is not an interrogation, so while you want to encourage sharing, you don't want to be perceived as pushy or intrusive. Avoid emotionally laden words that communicate your opinions because you don't want to prejudice the narrator's response. Rely on open-ended questions that do not telegraph what you think might be important; instead, simply establish the area that you're interested in, namely, the "territory to be explored while allowing the participant to take any direction he or she wants" (Seidman, 2006, p. 84).

This approach allows the participants to select and report what seems most important and meaningful to them. You can guide the conversation through follow-up questions that key in to the points that connect to your research needs. This approach allows you to build a contextual background so that you understand more of what you are hearing from your narrator's perspective. Charles Morrissey's (1987) enlightening article on the structure of context-based questioning gives an excellent introduction to the use of the "two-sentence format" for oral history interviewing, with the first sentence to establish the context and the second to request the narrator's input.

For example, assume that you have told your narrator, the mother of three school-age children, that you are researching the parents' perspective about educational services for children with special needs because you want to help policy makers know how their decisions affect children and families. You might start with a simple request that she tell you a little about herself and her family. From her general introduction, you could next ask her about her children and then move on to her experiences in parenting and accessing services for a child with special needs. Consider the following questions: *You mentioned that you have three children in school, right? Can you tell me a little about your kids?* And next, *You said that your youngest son, Billy, has autism. Talk to me a little bit about his experiences in school.*

By using open-ended requests, you extend an invitation to your narrator to sort through her memories and talk freely about whatever she considers relevant to your question. From her early response, which includes the fact that her youngest has autism, you invite her to tell you about Billy's experiences in the local school system and her experiences as the mother of an 8-year-old with special needs. You might ask her to compare her experience with the school regarding her two other children who do not require special services. Notice that these questions or requests for reflection build on context. By establishing a starting point for your open-ended question, you are telling your narrators what interests you and what you would like them to talk about.

In contrast to the context-based questions, consider this string of closed questions: *Do you have any children? How many? What are their names? Do any of them have a condition that requires special services? What is the condition?* You can see that this approach gives the interview the feel of an interrogation. How much more pleasant and comfortable it is to simply invite this information to flow naturally from a mother's conversation about her children. As she shares her story in her own way, you learn a great deal by observing how she introduces Billy into the discussion, and this path to discovery is more informative than what you might learn from closed questions that are more appropriate to survey research.

Table 6.1 gives other examples of open-ended versus closed questions for interviewing in connection to hypothetical research related to improving literacy programs. Notice how the first sample question only asks for one- or two-word answers, information that could more easily be collected through a review of school records or a simple survey. Sample question #2 takes a different approach, opening the discussion by confirming that the narrator had been enrolled in the program and then inviting him to talk about his experience. The researcher can then follow up on information that is provided in the initial response to learn more about the experience. As in this example, closed or structured questions that limit the possible responses are characteristic of the type found on a true-false or short-answer "objective" test. These are more characteristic of a questionnaire or survey. An open-ended question, which allows greater latitude in selecting details and refinements for the response, is more of the nature of an essay question. An in-depth interview relies on the personal *essay* to access meaningful insights and thoughtful reflection.

Table 6.1   Types of questions—open or closed?

Purpose for doing the research: To provide school administrators with the information they need in order to plan for successful, high-quality literacy programs for their students.
Research questions: What are the experiences of students participating in literacy programs? What are the benefits and what are the impediments to participation? How might the programs be improved?

Sample questions	Type of question (open-ended or closed)	Comments
1. Did you participate in the literacy program at Centerville? How many years were you in the program?	These are closed questions that don't invite conversation and limit the range of response.	To collect this sort of information, a records review or survey might be more efficient than an in-depth interview.
2. Mr. Chang tells me that you were in the literacy program. Is that correct? I'd like for you to tell me about your experience in that program.	First part is a closed question (yes/no) but it leads to an open question that invites participant to reflect on experience and share what personally seems most significant.	Response to this question could offer a productive point of departure for further discussion.

Skillful interviewing also requires that you consider the likelihood that your narrator will be able to answer your question. In an interview, you are asking for one person's view of a situation, and your questions should reflect this perspective. Continuing the example introduced in Table 6.1, assume that to help school administrators make decisions about adopting a literacy program for their district, you have chosen to focus specifically on experiential literacy programs. Since you want to know about social, behavioral, emotional, as well as academic outcomes, you have decided to interview students related to their participation in a program at a nearby high school. You intend to collect interview data from six to eight students and then assimilate, interpret, analyze, and conclude what works or doesn't work from the students' perspective.

Table 6.2 lists several questions that you are considering for your interviews (column 1), along with suggested revisions (column 2), and the reasons for the revisions (column 3). These examples demonstrate the need to ask questions that align with your narrators' range of experience and perspective and emphasize the importance of asking narrators about matters that they can discuss. If you want to learn something that narrators might not be able to answer, such as Question 1, ask for information that acknowledges their perspective. Similarly in Question 2, your narrators cannot tell you with accuracy how effective the program was, and indeed you will need to define for your research what you mean by *effectiveness*. If you are equating *effectiveness* with *academic progress*, then an in-depth interview might not be the best method for data collection. However, through an interview, you will be able to identify less quantifiable effects. Similarly, Question 3 asks the narrator to speak for others, instead of just for herself. Awareness of the narrator's perspective is matched by the importance of guarding against preconditioning the narrator's response. Questions 4–6 reveal that you feel the program has problems and that the student should share your opinion. These questions precondition the responses you will get, and you may never hear the positive things that the students may have to say.

You can invite additional discourse by following up on interesting points that your narrator has introduced. In initial questions, a researcher sets a topic or area of focus. In follow-up questions, the researcher lets the narrator know what is of interest and prompts discussion along the lines that will best advance the research. An interview is not simply asking a question—*What can you tell me about thus and so?*—and then making the narrator figure out what is or is not of interest to you or relevant to your research. "The best interviewers

Table 6.2 Asking questions from the narrator's perspective

Purpose for doing the research: To provide school administrators with the information they need in order to plan for successful, high-quality literacy programs for their students.

Research question: What are the experiences of students participating in experiential literacy programs? What are the benefits and what are the impediments to participation? How can the programs be improved?

Draft question	Suggested revision	Comments
1. What should the school board do to improve the literacy programs?	From your own experience, what can you tell teachers about how to make the literacy program better?	By framing the question from the student's point of view, you will be able to discern recommendations for the school board. The student may not know what the board should do, but can speak to what she would recommend to teachers.
2. How effective was the experiential literacy program in helping students develop literacy skills?	I am trying to learn more about the experiential literacy program at Centerville. Can you tell me about it?	Student has no way of knowing overall effectiveness and can only share from the personal perspective. In addition, academic effectiveness would best be researched through records review.
3. What do students think of the program?	What is your opinion of the program?	Student can only speak authoritatively of own opinion.

4. What do you think about having to waste time after school in this program?	Let's talk about your experience in the program.	Original question assumes that student sees program as a waste of time and turns the focus toward a negative response. It would be best to wait for the student to share own experience, noting how she views her participation.
5. Some students think Mrs. Dexter lets the older students pick on the younger ones. What do you think?	Tell me a little about how it feels to be in the class.	Original question biases student's response.
6. To be assigned to the program, you were identified as failing at least one class. What problems did this cause for you with the other students?	Did going to the experiential literacy program affect your relationships with other students? Can you tell me about that?	Again, original question biases student's response.

listen carefully between the lines of what is said for what the narrator is trying to get at and then have the presence of mind, sometimes the courage, to ask the hard questions" (Shopes, 2002, p. 3).

Many texts tell the researcher to "probe" deeply to uncover more information. From my own background in the humanities, *probing* connotes an almost clinical intrusion; it feels invasive and not at all empathetic. I prefer the concepts of *asking* for elaboration, *inviting* deeper discussion, or *encouraging* meaningful reflection and sharing, all of which set the stage for the narrators to explore and search their memories and consciousness. Whichever term you are comfortable with, the idea is to get beneath the first response you hear. This requires active listening as well as flexibility and patience.

Active listening

"Interviewers are listeners incarnate; your machines can record, but only you can listen" (Glesne, 1999, p. 81). After you ask a question for your narrator to consider, you are not just listening to the content of the answer. You're also listening for cues about how narrators relate your question to the larger story of their life memories. What emerges first may be the surface details, the public voice that "always reflects an awareness of the audience. It is not untrue; it is guarded. It is a voice that participants would use if they were talking to an audience of 300 in an auditorium" (Seidman, 2006, p. 78). Buried within the initial telling of the story, you find the markers or signposts that point the way to other promising areas to explore. Noticing these signs requires that you stay aware, attending to the message that lies beneath what is actually said.

Avoid selectively listening for a simple answer that forms a straight line to what you think confirms your assumptions. You are not seeking answers to prove a hypothesis or to corroborate what you think you already know. You are asking for *their* story and what they think is significant and meaningful in their own experience. Even though all of your preparation has readied you to research this ground of experience, memories and perceptions will vary greatly. The meanings that people ascribe to their experiences and the words they use to communicate them are equally diverse. The task, thus, is to appreciate the other's point of view and to listen carefully to the uniqueness of the information being shared. In following up on responses, you can pose questions or simple requests that will engage your narrator and deepen the interview experience. This process can be as simple as asking, *Can you give me an example of what you mean?* or *You mentioned that you had a similar experience earlier; can you tell me about that?* or *How did you solve that problem?* or *Tell me what you did when that happened.* Even if you think you know the answer, it is important to hear it in their own words. Attend to the multiple layers of language. There are words and phrases. There are pauses, body language, facial expressions, nervous laughter, deflections, nonverbals that allow "think time," and silence. Each word, act, and pause carries meaning. In your notes, jot down what you see as well as what you hear. A shrug or a simple sigh can communicate volumes. Understanding the context as well as the cultural or social norms can help you interpret some of what you see. For example, you will want to know whether the narrator keeps her eyes lowered because of embarrassment or whether it is related to a cultural norm. Capture on paper a simple note to remind you of what you observe in the way of emotion, hesitance, excitement, diffidence, and so forth, and then, after you have left the interview setting, return to your notes to fill in the essential details.

Remind your participants that your questions are limited by your perspective and that you would appreciate their consideration of what other topics should be explored. You might start the interview with this statement, and then repeat it as you close the interview. Next session, you may gain additional information and insights that you might not have thought to ask.

It is important to show your narrator that you are interested in what they are telling you. Be actively engaged in this process, but be careful not to signal enthusiastic agreement. Allow the narrator to tell the story; you listen and confirm that you have heard—not that you agree or feel the same. Your attention and indication of interest are what encourages and prompts the narrator to continue.

Notice what your narrators say and what they do not say. If they appear to avoid answering a question, you might ask the same question again differently. If it still isn't answered, realize that no answer is still an answer.

Sometimes it may not be clear exactly how their response connects to the question you asked, and you will need to ask them to help you understand what they have told you. For example, you might say something like, *When I asked you about your experience in high school, you told me that your brother has Down syndrome. Can you help me understand how that affected your high school years?*

Also be attentive to your own reactions to what the narrator is saying. Suspend disbelief and don't sit in judgment. If you start having internal disagreements with what your narrator is telling you, you may be operating from a position of judgment and your subjective lens may be getting in the way. Simply observe what you are feeling, but remember that you are there as a researcher, seeking information from sources outside of yourself. You are not there to engage in debate or to inform the narrator of another way of doing things. Similarly, you need to be alert to feelings of complete agreement with the narrator as well. When you find someone who thinks as you do, it might be more difficult to take the interview to a level where you can learn something new. Jot a quick note to yourself so that when you return to your home or office you reflect on your reactions and expand your thinking.

Ambiguity and contradictions often emerge in an interview. Don't attempt to confront the narrator about inconsistencies, but do express your confusion and see if he or she can clear things up. Sometimes it's simply a matter of a misstatement, sometimes it's a matter of forgetfulness, and sometimes it can reveal highly significant information. For example, in an interview, a narrator once gave me several different estimates about how long a time a particular event had taken. I didn't notice it during the interview, but it became quite apparent as I read the transcript. What I realized is that each time the event was discussed, different issues were being introduced: the greater the challenge that was faced, the longer the overall estimate of lapsed time. In this case, the perception of the element of time revealed a greater significance than actual clock time or days passed.

Flexibility and patience
The use of open-ended questions requires a great deal of flexibility on your part. Your interview guide will be helpful in getting things started, but with an open-ended question you really can't be certain exactly which direction the answer will take.

Be flexible in the sequence of your questions. You may have planned to cover a certain topic at the beginning of the interview, but find that an opportunity to explore a different topic presents itself at a time when your narrator seems interested in pursuing it. Don't derail that discussion; just follow the lead and remember your purpose and what you are attempting to learn. I have found that after starting an interview, I may not return to my interview guide until quite a bit later, and then will discover that much of what I had planned to ask has already been answered in other contexts. This simply demonstrates that people organize their thoughts differently. By allowing the conversation to flow in the sequence that makes sense to the narrator, you gain a more natural representation of his or her thoughts about things.

Allowing time for a well-considered reply may at first be a challenge. As difficult as it may seem to have your question met with silence, a delayed response may in fact be quite productive processing time. A too-quick response may signal a nonreflective answer, one that is given without much thought or consideration. Silence may signal that the narrator is considering how to answer your question or may need a moment to recall the information that you are asking for. Don't rush to cover the silence with the next question, but after an extended silence, you might try rephrasing your original question. If that still doesn't bring a response, move to another line of questioning and possibly return to this topic later, perhaps in another context, with the question posed in a different manner.

While we're on the subject of silence, be sure to allow several moments of time to pass after a narrator seems to have completed a thought. There may be more to come that you will miss if you rush into the next question. Don't disregard the silence, but notice it as one of those markers for consideration. It is also important to explore the gaps in the story as signals that map the world you are exploring (Norquay, 1999). There are messages and perhaps cultural significance in the silence of the unspoken word and in the cadence of memory. Silences and what appears to be forgotten or intentionally avoided can be as noteworthy as what is remembered and voiced. If you sense that your narrator does not intend to respond to your question and you choose to transition to another line of inquiry, first ask the narrator if there's anything to add and explain that you now would like to move to a different topic. Record this bit of information in your notes.

Patience is required in allowing silence to resolve, yet you will face another dynamic that requires patience at a potentially more challenging level. This is the patience for the telling of stories and sharing of reminiscences that may seem totally off the topic and without a point. You know that you have limited time with each narrator, and the temptation might be to interrupt a rambling tale that seems disconnected and irrelevant. However, I encourage you to appreciate the value of the story that may meander and appear to move aimlessly away from the question you asked. There was something in your original question that prompted this "bird walk" away from the topic. Listen carefully to the story that is being told; there may be markers that signal rich territory to be explored, even if it doesn't seem to answer your question of the moment. Instead of going on automatic pilot and simply smiling tolerantly until you can interrupt, listen and assess what question is being answered. You can learn a great deal by following the diversion. At some point the story may return to the question you asked, making connections that you could not have anticipated. If not, you may need to remind your narrator of your original question and ask what the thinking was that prompted this particular story. For example, if you are interviewing a student about the experiential literacy program and the answer seems totally off track, try asking something like this: *You know this is quite interesting, and I'd like to understand what prompted this story. I had expected to hear your opinion about the general environment in the after-school component of the literacy program—your feelings about meeting in the basement for those sessions—and you've just shared with me a story about one of your friends who transferred to another school district. Can you help me make the connection to the literacy program?*

Concluding the interview

As the interviewer, you will keep track of the questions you need answered as well as the time on the clock. If you are nearing the end of your session and you haven't accomplished what you had intended, let your narrator know that the time is nearly up and that you need to be wrapping up for the day. Often, the narrator will offer to give you a little extra time, especially if he or she is about to begin an interesting story, so leave enough time in your schedule that you can be flexible. However, you don't want the interview to go on and on for hours. First, this can be an exhausting endeavor, and second, after a while, you'll lose focus and experience a sort of numbing effect. You can tactfully break for the day and schedule your next interview, recording in your notes where you stopped and possibly how you might like to begin next time.

Thank the narrator at the end of the interview on tape. Leave the tape running for a few moments as pleasantries are passed, then turn off the recorder. If the narrator thinks of something to add, ask if it's okay to turn the recorder back on, or set it back up, if you've already packed to go. Just be certain that your narrator knows that you are recording. It's unethical to be taping without the narrator's awareness. In addition, a day or so following an interview, especially if it concerned sensitive topics, it may be appropriate to phone your narrator to check in and repeat your thanks for helping you with your study. If all seems well, you should confirm the time and place for your next session.

After concluding an interview, that's when you really get busy. First, you need to attend to the comments and observations that you hastily jotted in your notes. As soon after the interview as possible, review all that you have written and elaborate on your quick notes so that when the time comes for further analysis, you won't be left struggling to remember what a cryptic remark meant.

You also need to transcribe the interview before your next session. (See Appendix A-3 for suggestions on using voice recognition software.) There are two reasons for this urgency in completing the transcriptions. First, you want to transcribe while the interview is fresh in your mind; and second, the transcription of one interview helps to inform your plans for the next. You will want to review the transcript against your interview guide and use your summary form to help you manage the interview data and prepare for the next session. During the interview, you have already been engaged in a sort of data interpretation and analysis, for that is what allows you to ask the follow-up questions and assess with comprehension what your narrator is telling you. After a session, recording your thoughts and reflections in a research journal or on an interview summary sheet allows you to keep an account of what you have observed and heard. This is all part of the process of preparing for your next interview or, in the case of your final session, for the next step toward interpretation.

Mistakes to Avoid

There really is no one *right* way of interviewing. There are, however, many ways of interviewing that can lessen the chances of success. Rather than discuss in detail all the things that you can do to weaken your interview, I have simply listed a number of the most common mechanical and process mistakes. Many of these stumbling blocks stem from problems in preparation as well as from missing the opportunity to explore deeply. These mistakes are actually fairly easy to guard against:

- Failing to perform a sound on check your recorder(s)
- Forgetting to bring backup batteries and tapes to the interview
- Failing to come to the interview fully prepared
- Failing to transcribe tapes before returning to the subsequent interview
- Losing track of time and running over the agreed-upon session length without permission
- Neglecting to manage your data and document the topics that have not been explored yet
- Neglecting to show interest in what the narrator has to say
- Interrupting the narrator and cutting off responses
- Confronting or arguing with the narrator
- Failing to attend to the narrator's physical and emotional comfort
- Failing to show empathy when appropriate

- Misinterpreting what the narrator is saying and not correcting the record
- Taking comments out of context
- Asking leading questions
- Failing to listen actively and missing opportunities to explore topics that the narrator introduces as important
- Asking multiple questions at the same time
- Concentrating on the tape or writing extensive notes while ignoring what the narrator is saying
- Failing to backtrack on what appears to be extraneous narration in order to make connection to the question that was asked
- Failing to recognize and explore gaps in the story
- Showing impatience instead of tolerating silence
- Not allowing the narrator sufficient "wait time" before speaking
- Failing to show appreciation for your narrator's contribution to your study.

Don't be discouraged by this lengthy list. In-depth interviewing is a wonderfully rewarding endeavor, and it only takes a little practice to become quite competent. To hone your skills, it would help to conduct a pilot study that involves interviewing individuals who have awareness of the issues you are researching but who will not be part of your final research sample. This will allow you to practice in-depth interviewing, further refine your questions, and build your confidence so that when you do start the actual interviews, it will all seem second nature.

WHAT'S NEXT?

While you are conducting each interview, you are actively listening to your narrator so that you can begin to understand the significance of what was said. Chapter 7 explains how to process the interview data to interpret and communicate this understanding to others. To accomplish this, you need to confirm that you have interpreted your narrator's words accurately. The following chapter provides guidelines for working with the data, namely crafting excerpted narratives, completing a narrator check, analyzing the data, and reporting your results.

CHAPTER 7

LEARNING FROM THE DATA

CHAPTER TOPICS:

- interpretation and display
- narrator check
- cross-narrator analysis
- reporting of results

The gateway approach frames investigations in ways that are personal and accessible. What more intuitive way is there to learn from another than to ask a question and then to listen to the answer from that person's perspective. After completing the interviews, your task is to take what you have heard and then interpret, check for understanding, analyze, and report in a manner that maintains the integrity of the interview and honors the narrator as the source of information and insight.

INTERPRETIVE DISPLAY

An in-depth interview utilizing an unstructured style characteristic of oral history opens the door for meaningful discourse and reflection. The stories that are told, the emotions that are revealed, and the opinions that are expressed are significant to the narrator, and you, as the researcher, are charged with discerning the relevance of what has been shared with regard to your guiding research questions. This function requires insightful interpretation that maintains the contextual basis within which the stories are housed and necessitates an effective strategy for communicating what is learned in ways that connect to your intended audience.

The quickest way to reduce interview data to a manageable size would be to merely summarize what you have heard and synthesize what you think it means. This strategy, however expedient it may seem, risks losing the intricacies and nuances of understanding by negating the authority of the voice that spoke the words and gave them life. For gateway research to accurately reveal the salience of an interview series, the narrator's voice cannot be silenced; the interview data must be accessible and clear, directly connected to the narrator's understanding of the experience instead of dimmed by a researcher's filtered view.

Your acquisition of contextual knowledge and appreciation for the subtleties of the experience has prepared you to interpret and understand the significance of your data. After reviewing the transcripts to interpret what was said and what it meant in context, you can use techniques from poetic transcription to create an excerpted narrative to tell the story that helps answer your research questions. This is an important step in that it provides you with a tangible statement of your understanding that can be confirmed by your narrator as reflecting his or her intended meaning. This process also focuses the interview data and preserves the essential meaning for later analysis.

Similar to Seidman's (2006) use of excerpted quotations displayed in paragraph form, the purpose of an excerpted data display is not to cut and paste intriguing quotes, but to preserve the meaning and reflect the personhood of the speaker. This requires attention throughout the research process, not just at the end when you are working with the data. The successful gateway researcher is one who listens deeply, interprets accurately, discerns perceptively, and selects wisely the expressions to display for analysis. The goal is to master the narrator's meaning and then bring the raw data from the interviews to life so that they emerge from the page to embody human experience with multidimensional vitality.

The core process for completing this form of excerpted transcription is a fairly straightforward one that I learned about during my own search for a means to process data for the Columbine study. I was motivated by a desire to overcome possible problems arising from my subjective knowing of the inside experience as well as by the practical need to reduce volumes of data for analysis. With the help of Miles and Huberman (1994) who directed me to the work of Richardson (1992), which led me to Glesne (1997), I found a way to overcome the real possibility that my own experience as an insider would subvert my ability to research from the perspectives of others and to share their experiences in their own voice, not just retold through mine.

In describing the method she used for poetic transcription, Glesne (1997) explains that she began by coding and sorting the data into themes and patterns, rereading her narrator's transcript excerpts under one theme and "trying to understand the essence of what [the narrator] was trying to say . . . trying to make sense of the data but also attempting to use her words to convey the emotions that the interviews evoked in me" (p. 206).

A critical step in this practice is a skillful and informed review of the transcripts to sort data and identify patterns representative of each individual's experience. Your background knowledge and the theories or concepts that have informed your research play a supporting role in this process, but rather than starting out to find expressions or examples to prove a preconceived idea or to advance a theory, your task is to connect directly with the experience described by your narrator. Your prior knowledge should inform yet not precondition your interpretation. You may need to read a narrator's transcripts many times, noting emerging patterns or different points of interest each time. The more familiar you become with your narrator's story, the more faithful you can be in its representation.

Start by working with the transcripts for a single narrator. Once you have crafted a narrative for that individual, turn to the next one. This is the interpretive step; you are attempting to interpret and understand what each narrator has individually said in order to deepen your perceptivity for the subsequent cross-case analysis, which will look at what can be learned from all of your narrators.

The way that I envision this form of data reduction is not to focus on aspirations of poetic quality, but, basically, to reduce the data so that the message and the heart of the interview can be communicated. It is fine if you are a gifted poet, but that is not a prerequisite skill. This task is to interpret and excerpt the raw data for analysis so that you can come to conclusions that are grounded in the words of your narrators. Consistent with Eisner's dimension of description, the construction of an excerpted narrative attends to qualities of the experience and the narrator's perception of it. A well-crafted narrative provides a richly tapestried account that enables a sort of vicarious participation so that appreciation for the circumstances becomes possible.

The following list of steps describes the process of creating a narrative to serve as an interpretive display of your interview data. After reviewing the steps, consider the example that follows, which demonstrates the process in action:

- Read and reread all transcripts for a *single* narrator in their entirety.
- Review your notes from the interview sessions and your research journal to see how they might contribute to your understanding of the narrator's perceptions about the experience.
- Each time you read the transcripts, mark areas that seem especially pertinent to your questions. As difficult as it will be, some wonderful stories will have to be set aside for the time being, if they don't address your immediate research questions and purpose. However, mark these passages for consideration later.
- Highlight passages that are of value in communicating the story that is necessary to provide the context.
- Using the electronic version of the transcripts, cut and paste the chunks of relevant data into a single document.
- Examine these chunks, observing repeated phrases, patterns or recurring elements, sequenced stories, descriptive metaphors, and so forth.
- Go back through the material, settling on a tentative plan for sequencing the data in ways that will communicate the overall experience meaningfully. Some data will lend itself to being arranged chronologically. Some will be more thematic.
- Go back through the data again; highlight words or phrases that seem to be the most descriptive or moving.
- Pay special attention to phrases that are repeated verbatim; often they will form a sort of refrain that gets at the heart of the participant's relationship to the experience and may help you discern underlying patterns.
- Reflect on the patterns that you see and carefully consider broader themes that these might embody.
- Closely assess other words or phrases to determine if they are essential to communicating the meaning. Transition words, filler words, asides, etc., can be easily deleted at this stage. Remember, you will be using fragments, not complete sentences in your display.
- Arrange these fragments in a "string" that runs down the page, looking much like a poem (or a grocery list!) instead of a paragraph.
- Review again and again, each time becoming more focused, distilling the interview into its essence, the simplest, purest form that communicates coherently. Your goal is to evoke the experience and bring it to life, not describe it or summarize it from a safe distance.

- Keep in mind your research questions and your purpose. Readers need to know how it felt and what it meant to have this experience if they are to deepen their understanding of the findings that you generate through data analysis.
- Cut, delete, purge. There is power in the fewest words—select the words that allow the reader to connect at an elemental level into the experience and evoke an empathetic resonance of understanding. Pare it down to the *only* words—those words that are critical to communicating the essence of the narrator's experience and response.
- Give yourself permission to change verb forms, if necessitated by re-sequencing the excerpts, and to replace nouns with pronouns (and vice versa) to smooth out the rendering.
- Narratives can be sequenced chronologically or thematically. You can later decide how to present your data in the final report.
- Play with the placement of the individual passages. Switching the order of phrases might add dramatic effect. Just be certain that you don't change the meaning by doing so.
- Check the original transcripts one more time to be sure that there's nothing you left out that connects to what is emerging from your data.
- When you have prepared the narrative, repeat the process for each of your study participants in turn.
- Later, for data analysis, you will consider each excerpted narrative with respect to its message (a vertical consideration) as well as in regard to all of the other narratives (a horizontal analysis) to discern common themes, patterns, understandings, or differences that emerge across your study population.

There is no single right or correct narrative that can be excerpted from any transcript. It is certainly not your narrator's only story or the one and only expression of the experience. Your excerpted narrative may not be the one that the narrator would craft. It may not be the one that other researchers would derive. Selected elements may differ and the "word reduction" may produce different expressions. You may choose to sequence the excerpts chronologically, and someone else might choose to arrange them by patterns or themes. For your final report, you may settle on telling each narrator's story in its entirety, and someone else may break the narratives into segments or themes for discussion.

Regardless of the variety of expression, what is created, however arranged or reduced, must be an accurate statement of the narrator's experience in relation to the research questions. It must offer the elements of the story delivered through the narrator's own words, provide the content that helps answer the questions that drive the research, and communicate in ways that foster a deep understanding of the experience. You may choose to repeat powerful phrases, creating a sort of poetic refrain, or you may elect to simply present what you see, without attempting a poetic style.

Presented with the original expressions intact, a clarity emerges from the narrative since it *recreates* the experience instead of *telling about* it. Data are enriched and enlivened through condensed expression by using the *only* words, creating verifiable intensity, evoking connection, and deepening comprehension. The content could be delivered with prose, but the white space around the phrases, the trait that makes it look like poetry, focuses the reading and directs attention to each phrase, making it stand alone and demanding to be considered.

Reducing interview data in this way allows study participants to share their own individual story, making it less likely that a researcher could interject personal opinions as one of theirs. The narrator check (see following section) provides a way to confirm this accuracy.

To demonstrate the process of creating an excerpted narrative, I offer the following example taken from an inquiry that I completed with mid-career doctoral students. My research questions had to do with the doctoral experience, what motivated students' decision to pursue an advanced degree, the challenges they faced, the supports that helped them through, and the advice they would give to others.

The first example (Exhibit 7.1) provides a few brief excerpts from a 21-page transcript of an interview with Elizabeth, who was completing her doctorate in education. After I had transcribed the interview, I went back through the text many times, to check it against the audiotape for accuracy, and to be certain that I understood the essence of the story that was being told. I first interpreted what I thought she was telling me, noted patterns in her interview, and relied on the research questions as a primary guide to help me identify and highlight passages that I thought seemed most effective in communicating her story. Some of the phrases I identified conveyed the details or facts of Elizabeth's experience. Some revealed her emotions about the situation, and some were important in conveying the overall plot or story behind her decision to seek a doctoral degree.

After highlighting the passages that seemed significant in Elizabeth's transcript (indicated in bold font here), I cut and pasted these excerpts

Exhibit 7.1   Excerpt from Elizabeth's interview transcript, highlighting key phrases

CM:	I am interested in your decision to pursue a doctorate. Can you tell me a little bit about yourself and how you came to be doing what you're doing?
Elizabeth:	. . . We were either going to go back to California—the Bay area where my family is—or move to Denver. So we chose Denver. **I started working at a local school district right away—I was a literacy resource teacher** which is kind of like a building resource teacher, like a curriculum coach, **one of those kind of ancillary positions that support classrooms.** But the way this principal wanted to do it was to really support her, and **I was trained with a Masters in school administration,** and she didn't have any assistance, so I helped her mostly, **I did whatever I could to help** her. And then I went to another school and then I went to middle school and **was an assistant principal for six years—that was in _____ district. I was there for eight years.**

    I loved my job I was really good at it, I really loved it but I was getting really—I wasn't emotionally bored, **I was emotionally overwrought—**you know **it's just so intense with the kids and the parents and the teachers and all that it's just very exhausting—just exhausting work. But I was intellectually and academically just not challenged.**

    . . . There were lots of things to learn but it wasn't what I wanted to learn, number one, and it was more like learning, well here's a new software for student tracking. Or here's a new, district mandated something. **I wasn't learning anything that was enriching—**it wasn't content, no— I mean it wasn't . . . I had to do all of my professional learning by myself, on my own, because the way that, at that time, the situation was, AP's do all the work and the principals go to a lot of meetings. I was an AP for six years and I worked really hard—**10-, 12-hour days you know to keep it all rolling—**

CM:	So the literacy thing got sidetracked?
Elizabeth:	Well I was asked by the language arts department to do a lot of professional development, and I oversaw the library—so I do a lot around the literacy . . .
CM:	So you were doing multiple jobs—
Elizabeth:	**Oh, I had so many jobs I couldn't even tell you [laugh] basically I was responsible for a lot, had a lot to do. I had 450 kids and their parents under my wing.** I probably had 25 teachers that I was responsible for, and about four or five different departments—all of the clubs, all of the activities, all of the parties, all of the celebrations, and all of the testing—and so there was a lot. . . .

    Anyway, I had far earlier in my AP career thought that I would get my Ph.D. part-time, that I would go to work for a couple of years, learn that, get kind of wired, and then I'd start going to school. **But I was exhausted, there was no way that I could do it—**by the time I got home I was catatonic—Is that when your mouth sort of hangs open and you can't really say anything? [Laugh] I couldn't do anything. I thought I can't do this,—I had always wanted to do it, but I realize that I couldn't. **So after about five years—four to five years—I knew I needed to make a change.**

into a document that would serve as a starting point for drafting her narrative for my study (Exhibit 7.2):

After culling these excerpts, I next studied them to find the *only* words, those words that I considered so essential to the meaning that they could not be eliminated. This technique of data reduction using poetic transcription actually corresponds to the advice that my high school English teacher, Mr. Carr, used to give to his students, namely, assume that you have to pay your readers $10 for every word that they read. Reduce, reduce, reduce to save dollars by using *only* the most powerfully concise and exact words possible! The challenge is to distill to the essence, to cut to the minimum wording that will communicate the meaning embedded in the narrator's reflections.

When I had completed what I thought might be a final excerpted narrative for Elizabeth, I checked it against the original transcript to

Exhibit 7.2   Key passages excerpted from Elizabeth's transcript

---

I started working at a local school district right away –
I was a literacy resource teacher
one of those kind of ancillary positions that support classrooms.
I was trained with a Masters in school administration,
I did whatever I could to help
was an assistant principal for six years—that was in _____ district.
I loved my job I was really good at it,
I really loved it but I was getting really –
I was emotionally overwrought, it's just so intense
with the kids and the parents and the teachers and all
it's just very exhausting—just exhausting work.
But I was intellectually and academically just not challenged.
There were lots of things to learn but it wasn't what I wanted to learn.
I wasn't learning anything that was enriching
10-, 12-hour days you know to keep it all rolling—
I had so many jobs I couldn't even tell you [laugh] basically
I was responsible for a lot, had a lot to do.
I had 450 kids and their parents under my wing.
I probably had 25 teachers that I was responsible for, and about four or
    five different departments
all of the clubs, all of the activities, all of the parties, all of the
    celebrations, and all of the testing—
and so there was a lot.
Anyway, I had thought early in my AP career that I would get my Ph.D.
    part-time,
But I was exhausted, there was no way that I could do it—
So after about five years—four to five years—I knew I needed to make a
    change.

---

ensure the integrity of the creation. I next scheduled a narrator check, so she could check that I had correctly interpreted what she was telling me and had accurately presented her experience. During this discussion, Elizabeth expressed some concern about the draft I had created for her. She commented,

> I think you captured the essence of the exhaustion and intellectual desire for more so I would have more options in my work but I'm not sure I articulated that last point as clearly to you as I have to myself— that for me was the essence of the decision to get my Ph.D. I really needed to have a chance to think and understand what was happening and life at school was so intense—we were told to do so many things I couldn't think deeply, reflectively, or knowingly. I wanted options—as the AP I felt somewhat trapped in response mode.

With this clarification, I was able to correct and complete Elizabeth's narrative. I excerpted significant expressions from the transcript of this additional conversation and e-mailed Elizabeth the revised narrative for a final check. She was pleased to see that her story had been amended and now authentically represented what she had wanted to say (Exhibit 7.3). The resulting narrative serves as a gateway, pulling the reader into the busy and demanding life of an assistant principal

Exhibit 7. 3   Selection from Elizabeth's excerpted narrative

Working at a local school district.
I loved my job
was really good at it,
loved it but overwrought
just so intense
with the kids
and the parents
and the teachers—
just very exhausting,
exhausting work.
10-, 12-hour days to keep it rolling
450 kids and their parents under my wing.
probably 25 teachers,
four or five different departments,
all of the clubs,
all of the activities,
all of the parties,
all of the celebrations,
all of the testing—

(Continued)

Exhibit 7. 3 Continued

Intellectually and academically just not challenged
I wasn't learning anything enriching
After about four to five years—
I knew
I needed to make a change.
Life at school was so intense—
couldn't think deeply, reflectively, knowingly.
Needed to have a chance to think—
understand what was happening.
I wanted options
That for me was the essence—
the decision to get my Ph.D.

---

and giving access to situations and feelings that prompted her decision to enroll in a doctoral program.

In terms of basic content, Elizabeth's excerpted narrative could be summarized by saying something like, "The narrator was overworked and nearing burnout in her work as an AP. She enrolled in a Ph.D. program because she wanted to be intellectually challenged. She wanted options." However, Elizabeth's own voice is far more effective in communicating the reality of the demanding work and experience that motivated her to make a change.

Since many have asked me the reason for formatting the display to look like a poem, I'd like to demonstrate the advantage that is gained by displaying the excerpts in this form instead of in a paragraph structure. Consider the following rendition of the same material that appears in Elizabeth's narrative:

Working at a local school district . . . I loved my job . . . was really good at it . . . loved it but overwrought . . . just so intense . . . with the kids . . . and the parents . . . and the teachers . . . just very exhausting . . . exhausting work . . . 10-, 12-hour days to keep it rolling . . . 450 kids and their parents under my wing . . . probably 25 teachers . . . four or five different departments . . . all of the clubs . . . all of the activities . . . all of the parties . . . all of the celebrations . . . all of the testing . . . Intellectually and academically just not challenged . . . I wasn't learning anything enriching . . . After about four to five years . . . I knew . . . I needed to make a change . . . Life at school was so intense . . . couldn't think deeply, reflectively, knowingly . . . Needed to have a chance to think . . . understand what was happening . . . I wanted options . . . That for me was the essence . . . the decision to get my Ph.D.

The paragraph display seems to lose the vibrancy of the expression, flattening out emphasis into a visually homogenized display. Silences and pauses are obscured as are the subtle rhythms of repeated phrases. It is no easier to present data in the paragraph form than it is to construct a display in a poetic structure. There is little to gain yet a potential for powerful communication to lose.

The presentation of interview data as an excerpted display touches a literary chord that resonates with aesthetic expression. As much as I appreciate the rising interest among qualitative researchers in poetics, and am inspired by both Laurel Richardson and Corrine Glesne for their originality in bringing this medium into the research arena, for gateway, I prefer to call the display an *excerpted narrative* because there have just been too many questions asked by too many individuals who fear that they would need to become poets in order to use this technique. The point of the narrative is not to stand as poetry. It is purely a means to distill and display a narrator's words in a story-like or thematic presentation that informs a research question and provides a gateway to the analysis and understanding of an experience.

Displaying data in this form is visually powerful. Like poetry, it makes it possible to condense expression and focuses attention to reveal multiple levels of understanding at the same time, not just those that fit within a mental construct perhaps more easily expressed in paragraph form. Paragraphing requires attention to structure and grammatical elements, and the visual shape of a paragraph puts everything on an equal footing. A display that has the shape of a poem is not so restricted. Language of use, without concern for grammar or structure, concentrates the reader's attention directly on the message that is being conveyed, while the surrounding white space separates and clears away that sense that all words and phrases are of equal value deserving equal attention.

This approach is especially useful in capturing the experiences of those who may lack fluency in English or those who are among marginalized or special populations that are often overlooked in research. A narrator is not disadvantaged by a lack of mastery of the predominant language structures, conventions, or grammar. Only the critical or essential words are reproduced, without attention to complete sentences, verb agreement, or misplaced modifiers. This open and accepting model for representation and later analysis of spoken data can access and share the voices of narrators from diverse backgrounds and multiple perspectives. The voices of experience can be heard, and life is breathed into the words on the page.

Other examples of reducing transcripts to create this type of display are provided in the appendixes. Appendix C-1, which is taken from the same project as Elizabeth's, presents excerpts from interviews with Bill, a mid-career marketing executive who was completing his doctorate in business administration. Bill's example demonstrates how the discernment of patterns throughout the transcripts facilitates the construction of an excerpted narrative. A complete narrative from the Columbine study appears in Appendix C-2. This narrative, taken from the interviews of a mother whose daughter was a Columbine student at the time of the shootings, demonstrates the effectiveness of this manner of data display, reducing almost 40,000 words from the interview transcripts into about 3,750 words to reveal the human dimensions of the experience across a broad spectrum of topics.

Crafting narratives may not be a skill that you feel immediately comfortable with. You may need to practice this form of data reduction on other transcripts before you begin to work with the interview transcripts for your study participants. Go back to the transcripts from your pilot interviews, or interview colleagues about an interesting topic and practice reducing and developing narratives for those. The more you practice this form of data display, the more comfortable you will become.

NARRATOR CHECK

The work of an interviewer does not end when the final narrative is completed. Questions have been framed and answers to those questions have been given. The next step is to check back with the narrator to make sure that you have truly understood and accurately portrayed what he or she told you. An easy approach to this task would be to simply give the narrator a copy of the audiotape and the transcript and ask that it be checked for accuracy and completeness. Standard practice accomplishes this essential quality control, the process that Lincoln and Guba (1985) call a *member check*, in a fairly straightforward manner. "Copies of interview transcripts are returned and reviewed together by investigators and interviewees. The work of understanding the materials is a joint effort and understandings arrived at enter into planning and development of next stages of the study" (Mishler, 1991, p. 127). This is an important step in checking for accuracy of the data. There is, however, another aspect of confirming accuracy, and that is to ensure that the representation of the interview to the ultimate reader reflects the meaning that the narrator intended.

Research relies on the communication of data in a form that can be read and reflected on by the consumers of the information. With in-depth interviews, the vast amount of data to be considered makes this an almost overwhelming task. While some interview researchers develop profiles or vignettes from the interviews and others paraphrase, summarize, or graphically represent relevant data, the gateway process relies on the preparation of excerpted narratives from the narrator's exact words and phrases. Since preparing a narrative or any form of rendering requires interpreting the narrator's meaning, I feel that it is important to check not only the accuracy of the recording and transcript but also to confirm that the interpretation of its meaning is accurate. For my initial research, I returned the audiotapes and transcripts for confirmation. Afterward, I created narratives to communicate the narrators' stories to the reader and gave them to the narrators for a second check. I discovered, however, that I could have simply worked with the transcripts and returned those with the narratives, for, in fact, the participants did not listen to their tapes or examine their transcripts. They did, however, closely scrutinize their narratives.

It may seem risky, asking a narrator to confirm that you have understood correctly, but, in my mind, ethical practice demands that you do so. It is indeed their story. You need to base your analysis on an accurate understanding of the data. To do otherwise would risk misrepresenting your narrators and undermining the integrity of your work.

In the process of creating narratives from interview data, the researcher's knowledge, purposes, and personal experience serve as filters for the data that will be used in the analysis. It is impossible to remove the researcher from the research process, but it is possible to strive for authenticity and to ensure that the presentation is consistent with the narrator's sense of story integrity. This is the point and the purpose of the narrator check.

Confirming the accuracy of your data display verifies your interpretation of what you have learned from the interview. For a gateway study, a narrator check provides an opportunity for a final interview, one in which the narrator interacts with the narrative and considers how the meaning that was interpreted by the researcher may or may not be accurate and complete. It is also an opportunity for the narrators to reflect on deepening understandings that might have emerged in the interviews by reading how their words are understood by the researcher.

To complete this process, give each narrator a copy of his or her interview transcripts and the narrative that you have drafted. Set a convenient time to discuss the narrative, making sure to allow the narrator time to review the material before you meet. At the narrator check, tell your narrator that your job as a researcher is to ensure that his or her voice is accurately heard, and that you need help in doing that. Be sure to tape-record this session, since it will be contributing additional data to your research. Ask your narrator to consider the narrative that you created and to make sure that you have understood and represented the intended meaning accurately and completely. Since your rendering required you to interpret what you thought they said and meant, this step is critical in confirming the validity and dependability of your data. Ask what they noted in the narrative and if there were any surprises. Encourage reflection on any points that are brought up.

For the narrator check, your focus is on the individual. You are checking the individual narratives against that narrator's story and your understanding of it—the patterns that emerged for each, not the patterns that may emerge from the entire sample, that is, not the analysis of all narratives. Please note that at this point, you have not confirmed the accuracy of the individual stories of everyone in your study so you are not yet able to share the narratives of the other participants. You are attempting to get confirmation from your participants that your presentation of the data for them is correct. You are not asking for agreement that your findings are correct, since you have not yet begun your analysis and since analysis requires the informed scrutiny as well as a conceptual (or theoretical) background to reach defensible conclusions.

Validating the faithful representation of draft narratives (with changes or corrections that are indicated) allows the refinement of your preliminary interpretation. Asking the narrators to evaluate the accuracy of the narratives that you constructed can confirm that the narratives capture their meaning accurately or, alternately, can allow an opportunity to correct the account. In any instance where you have misinterpreted the meaning, the narrator check makes possible additional conversation so that meaning can be clarified.

Narrator checks of this sort can also provide access to rich data that were not made available during the original interviews. It becomes possible to reflect on the reflections. Narrators often observe patterns in their own experience that they had not been aware of previously. With the opportunity to see how their words are being understood, narrators can also see how their words were misunderstood, or how sarcasm was apparent, or when key points that they had wanted to make were overlooked. For example, when the narrator check with Elizabeth revealed a problem, I was able to amend her narrative to incorporate additional details. Upon seeing the corrected narrative, her comment confirmed the importance of this step: "Yes, I like that narrative much better! It says what I meant. Narrator checking is really powerful."

In the Columbine study, the narrator check served an additional purpose. That study was considered a sensitive investigation requiring special measures for protecting privacy and concealing identity of the participants. The narrator check provided an opportunity for participants to check for accuracy and completeness and, just as important, to scrutinize the narratives so that any identifying information could be removed. Participants were also asked to share their narratives with family members to make sure that they were comfortable with the details that were being shared.

Sophia, whose excerpt you read in Chapter 4, commented in her narrator check that she had not previously recognized the patterns she was now able to see in her narrative. By studying her own words, and the way that I had understood what she had said, she realized that she was given a chance to evaluate her own interview information just as I was. She observed that this opportunity for expanding awareness could lead narrators to become their own change agents, and that once they could see how situations in their life had shaped their experience, they might be better able to take action to make things better for themselves and for others.

In addition to allowing your narrator to inform your understanding of the interview, a narrator check gives you an opportunity to confront your own subjectivity and to face your biases head on. On the rare chance that your narrator wholeheartedly disagrees with your presentation of the interview, consider what dynamics are at work. Has something conditioned your misunderstandings or tainted your rendering of the data? Perhaps you disagree so completely with what the individual said that you were not able to hear from the narrator's point of view. A situation such as this will require some soul-searching, and probably conversations with your advisor, to get to the heart of the matter. Whatever the case, you started the research with a commitment to research from the narrator's point of view, and you will need to reconcile this dilemma before moving to the analysis stage. The power of this type of accuracy check is that it acknowledges the ultimate authority of narrators with regard to their perceptions about their own experience.

ANALYSIS

While a great deal can be learned from each of the excerpted narratives, larger patterns and themes can be discerned by considering all of the narratives in relation to each other. This moves into the analysis phase, the point in the process where you go beyond the interpretation or comprehension of an individual's intended message into a thorough analysis of what the data mean across all narrators. Single narrators have demonstrated individual patterns of response; if you have included more than one narrator in your study, for analysis, you need to reflect on how all of the individual patterns fit together, overlay, coincide, or contradict. You examine these patterns and their interrelatedness through the lens of the frameworks or theories that you identified for your study.

When you started your research, you verbalized your purpose and the questions you wanted to answer. To analyze the excerpted narratives, you now need to return to your point of origin and determine how you can contribute potential answers that deepen understanding with regard to these starting points. To accomplish this elemental function, you rely on research in the field, your increasing understanding and expertise, the patterns you have discerned, and your knowledge of the use to which the research will be put.

There are many strategies for undertaking this step, and in truth the term *analysis* is used to mean different things in different research traditions. An educational criticism, for example, relies on interpretation to explain the meaning; evaluation to determine if what has been observed has fostered growth, arrested growth, or produced no effect whatsoever; and thematics to disclose common patterns that have relevance elsewhere. As a result, a sort of evaluative appraisal supplants analysis. Analysis in oral history looks for "plot, key phrases, structure of the narrative, context of the life, self-concept, contradictions, omissions, choices, desires, metaphors, symbols, and the influence of the individual's work" (Yow, 2005, p. 307). For gateway research, analysis may involve a combination of any or all of these foci, yet the core requisite is that it includes a scrutiny of the patterns that are discerned across the narratives and a critical appraisal of their significance and their implications when viewed in relation to the knowledge base in the field.

Analysis will probably not be as neat and tidy as the process I am about to describe for you. You may go through the individual narratives many times, perhaps noticing different themes, codes, or patterns than you did when you first created them, and then struggling to make sense of how that adds to or contradicts your understanding of the data. However, for the sake of brevity, let's assume that you complete this stage in a straight-line fashion.

You moved toward your analysis when you interpreted and excerpted the data for a descriptive narrative for each of your study participants. The narrator checks that confirmed you had correctly represented the meaning allow you to begin a cross-case analysis, which means that you will now look at all of the patterns that have emerged in the narratives of all of your participants. You are looking to make connections across the narratives, noting the commonalities of experience and response as well as the differences, and then accounting for and making sense of what you see.

At this point, consider each narrative with specific regard to each of your research questions. One way to do this is to create a separate document for each question and then review the narrative that you created for each narrator to highlight the sections (themes, patterns, or discrete elements) that appear to be related to that question. Make certain that you don't lose sight of where you take the data from. Exhibit 7.4 provides one schema for managing the data for a single narrator. These excerpts were selected from the narrative of a student, David Martinez, who told of the problems he had with the scheduling of after-school literacy classes.

Exhibit 7.4 Model for recording patterns from narratives

Narrator: David Martinez
Project: Experiential Literacy Program
Date: 4/09
Challenge: Problem with scheduling

Narrative p. 3:
> Hated it
> Didn't want to stay late for the stupid class
> School was over
> but we had to stay
> and listen
> and try to learn
> I wanted to leave
> Just wanted to go home

Narrative p. 5
> Lots of absences
> we'd all just ditch
> friends are busting out at the end of the day
> but we were left slumping down the hall
> No wonder kids ditch class
> it's time to be going home
> we wanted to go home too.

Narrative p. 14
> I hated it because I had to stay late –
> and not get home until after dark

Narrative p. 18
> the other thing—
> I wanted to be in the tech club
> No, couldn't do that
> had to go to that stupid class
> felt like a dummy

As tentative patterns and thematic schemata emerge, take what you find in an individual narrative and develop a master list, reviewing the selections that you have identified and then returning to the original transcripts to take special note of the patterns that you have found in other narratives. This will help you determine if you have overlooked a common theme within other narrators' raw data. You may find it helpful to develop a sort of matrix for keeping track of the themes or patterns that you see emerging from your excerpted narratives. A simple grid with the names of your narrators along the top and then the common themes along the left margin (Table 7.1) allows you to organize your observations about what the narratives reveal. A template for this type of display is included in Appendix B-5.

Use your guiding framework and background knowledge to help you discern the implications of the patterns that you are seeing from your data. Are the experiences, reactions, and understandings consistent with what prior research has found? How does earlier research inform and help you understand what you are seeing? How does your research contribute to the body of knowledge? Are there "outliers" in your study that offer disconfirming perceptions? What do you think that these might signify?

The excerpted narratives are your data for analysis, not exclusively the product of your research. They are not intended to categorize the findings or list the recommendations. The purpose of the individual narratives is to serve as the data display, the contextual reminder of the basis for your conclusions. For example, having asked your narrators to talk about their experience and to tell you what advice they would have given themselves (or to others in a similar situation), you have a

Table 7.1   Data analysis organizer

Narrators / Themes	Paula Rosen	David Martinez	Melena Sharone	Nicholas McGuire
Schedule difficulty	X	X		
Strong curriculum	X		X	X
Problems with class size		X		X
Positive peer support			X	X
Issues with physical facility	X	X	X	X

starting point from which to analyze significant factors connected to the situation that may have implications elsewhere. In the Columbine study, two of the narrators said that they would advise others not to let relationships with their spouse suffer by focusing exclusively on the needs of the child. This comment contributed to a finding that is significant not only to individuals exposed to trauma but also to service providers who may be charged with planning for crisis recovery or family counseling in a grief-stricken community.

In the previous section on crafting an interpretive display, I noted that you may have to bypass interesting stories that do not connect to your research questions. When you've completed your analysis of the data that relate to your questions, return to the transcripts and see if you can identify a pattern in those "off-topic" sections. If they don't answer your questions, what questions do they answer? Most of the time, these asides are not really productive in a research sense, but in the chance that your narrator has provided insight that applies in a broader way to your field of investigation, it is worth taking a look and deciding if you find something that should be included in your report as an addendum, or perhaps something that gives you direction for future research. If you intend to add the information to your study, be sure to confirm with your narrator that you understand it correctly.

It is important to check for your own bias throughout the research, but pay special attention during the analysis stage. Ask yourself again what is contributing to your awareness of certain patterns. Return to your research journal to remind yourself of tendencies to think subjectively. Consider how your own experience has conditioned you to pick up on certain points to the exclusion of others. Once again, commit to being your own devil's advocate by looking for alternative understandings and asking quite literally, "what's wrong with this picture?" Look at the patterns you identify and the significance that you find from a different perspective. Your narrators might conclude something different, and your advisor and other researchers might disagree with you. Remember, interview research is often messy, ambiguous, and contradictory. This is the point at which you challenge yourself to reach defensible conclusions and to provide your reader with your rationale and justification for those conclusions.

After you have discerned patterns in the narratives, apply a different, wider lens to assess what the data are telling you. Look for the broad headings under which the patterns might be grouped. For example, perhaps you find that your narrators relate similar accounts of their difficulty in accessing services, problems in getting essential information, and the failure of decision makers to understand the problems that are faced. After addressing these patterns, consider the broader frame under which they might be categorized. In this case, the importance of clear communication emerges. The consolidation of these patterns into broader themes allows you to summarize implications for others in similar situations.

As you close out your analysis, you are probably also aware of the limitations of your study. You have shed light on complex social or educational situations and experiences. In concluding what it means, you also need to conclude what it doesn't mean and point to additional inquiry that might help expand understanding of the topic.

REPORTING

After analysis, it is time to report on your research. A gateway study, as with any qualitative investigation, is an attempt to add to what we know about the world, to inform matters that have yet to be illuminated fully, and to pursue answers to questions that can shape a deeper understanding about some aspect of experience. "Qualitative research investigates the poorly understood territories of human interaction. Like scientists who seek to identify and understand the biological and geological processes that create the patterns of a physical landscape, qualitative researchers seek to describe and understand the processes that create the patterns of the human terrain" (Glesne, 1999, p. 193).

Research has value only when it is shared in ways that communicate what is learned to others. But with any qualitative inquiry, you are not generating a single, irrefutable answer to the question you have asked. You are, instead, sharing with others what you have learned from your narrators about how they have experienced and take meaning from some aspect of their life. Since the diversity of human response at times seems almost endless, all that can accurately be reported is that among the individuals you have interviewed, certain patterns of response can be discerned. The stories you have heard do not represent the full range of experience and response to the situation you are researching. To summarize your study findings without acknowledging this fact risks erasing the uniqueness of the individuals involved in your sample and elsewhere. You cannot draw conclusions about the whole of the experience, but you can contribute a deeper understanding of patterns among your narrators related to at least a part of it. Awareness of these patterns helps inform others about that circumstance or situation, and this knowledge has utility.

When you have exhausted your analysis and you can verbalize what you see from the data, you will turn your attention to reporting on your research in ways that convey the patterns you have observed within the context you were investigating. You can decide whether you want to present the excerpted narratives in their complete form, that is, one narrative for each narrator, or whether you prefer to arrange your data for discussion within sections for each of the patterns or the themes that you have identified. This is your research, and you will decide how best to report it. The overarching goal, though, is to keep the narrator's voice present and make your style of writing accessible. Respect for that voice is what allowed you to collect, interpret, distill, and analyze your interviews. Ultimately, it is the voices of your narrators that create a gateway to deeper understanding of the experience. However, it is important to tell your reader that the narratives do not represent direct quotations as written, but have been excerpted from the narrator's words and phrases and then sequenced into a readable form for presentation.

The purpose of the research and the research question will determine the most effective strategy for presentation. For the mid-career doctoral student inquiry, I chose to arrange the presentation thematically, first discussing each pattern I observed and then providing selected aspects from the narratives to demonstrate those patterns. Other inquiries I have presented with the data arranged chronologically to maintain the sense of story.

For the Columbine study, in order to provide a sense of the individual whose experiences were being reported, I wrote an introduction for each narrator (see Appendix C-2) and then presented the excerpted narrative. I felt this presentation of the story as a whole helped create an awareness of the human quality of the total experience. After displaying all of the narratives, I discussed my analysis under the headings of my specific research questions (challenges, responses, resources, and recommendations), using the evidence displayed in the narratives to support the analysis. Following the analysis by research question, I presented conclusions related to the broad themes that were revealed in the data, namely the importance of location (sense of place), the perceived intention of service providers, and the need for connection and reconnection.

While I felt that the analysis had answered my research questions, I was aware that the primary question, "What is the experience of parents whose children are exposed to a school shooting?" was not answered holistically. Granted, I had reported specifics that allowed me to make recommendations for other communities faced with trauma and loss, but the question about overall experience had not been answered to my satisfaction. To achieve this purpose, I returned to the narratives of the individual narrators and created a collective narrative, a mosaic that juxtaposes selected excerpts from all of the narratives into one document. A mosaic creates a sort of *gestalt* for the collective voice of experience. The process for completing such a display is essentially identical to the process of creating a narrative from the transcripts for a single narrator. The only difference here is that instead of excerpting from a single transcript, phrases are pulled from the narratives of all of the study participants to embody the overall experience that they have expressed. Exhibit 7.5 is an excerpt taken from the *Columbine Mosaic*. The complete mosaic is provided in Appendix C-3.

In your research, when you proceed from the individual voice (the excerpted narrative) to the combined voices (the analysis), I encourage you to consider if it might be appropriate to create this type of final narrative to capture the shared experience. A mosaic presents the collective voices, representing the varied and multiple lenses that have been shared and serving as a reminder that there is no one way to perceive a situation or event. Your mosaic may reveal a harmony of understanding, a confusion of differences, or a collage of fractured expression. These should preserve and present the authenticity of individual differences in a collective narration.

Exhibit 7.5   *Columbine Mosaic* (excerpts)

. . . Shooting
Children slaughtered
Kids flying down the hall
Teachers got them out
Stay down, stay down!
Run!
Sniper on the roof
Moments of terror
Kids in the library
What safer place?
God, it was raw
Emotionally nothing left
Vulnerable
Empty
Guilty for surviving
Guilty for finding my child so soon
Bittersweet
. . . . .
If it could happen here
It could happen anywhere.

To create a mosaic, review the excerpted narratives that you have written for your individual narrators. Read through them to highlight unique or particularly evocative expressions that reveal the diversity and multiplicity you have discovered. Look for the metaphors or symbols that provide insight into the diverse ways that meaning is made. In addition to rounding out your research, a narrative provides a group or an individual over time with powerful perspective because it provides a global view that transcends their specific situation. A mosaic takes understanding beyond the limits of knowing that a prose summary or conclusion could provide. Reporting your research with vivid display focuses awareness and demands attention. Ultimately, it makes real the circumstances and complexities that deserve to be known so that recommendations or conclusions can be drawn.

Through the gateway process, you are not *giving* a voice to your participants; you are *honoring* their voice by giving it a medium through which to be heard. These voices can speak, through your study, to educators, policy makers, practitioners, and others so that situations can be better understood and, if necessary, conditions improved. Recurring themes that emanate from the stories you are sharing can transcend the specifics of the narrators' experience and contribute to a fundamental awareness that could have significance elsewhere, perhaps enabling practical solutions to be developed. Frame your discussion with these potentials in mind.

In your final reporting, consider how future studies can build on what you have discovered and answer those questions that you stumbled upon in the process of your work. Suggest how other investigators might extend your work, perhaps investigating different perspectives related to that experience, utilizing different lenses, different narrators, different aspects of that community of experience.

WHAT'S NEXT?

The purposes and design of in-depth interview research have been explored, processes of the gateway approach examined, ethics reflected on, and techniques considered. Next, we weigh the implications for using the gateway approach as a means to learn from the experiences of others. The following chapter addresses possible research environments and applications for the approach, and, since I first used the gateway model for research into a life-changing event, it offers points to remember should your research lead you into a study requiring special sensitivity and care.

CHAPTER 8

THROUGH THE GATEWAY

CHAPTER TOPICS:

- horizons, implications for future studies
- research in varied disciplines and environments
- special cautions for sensitive research

Learning from the life experience of another is a privilege, and while connecting across boundaries of understanding can be a soulful endeavor that privileges us all, it is not easy work. An ethical purpose to drive the research, a tolerance for ambiguity and uncertainty, a willingness to explore unfamiliar and perhaps shadowed terrain, and a commitment to the highest standards of research practice will be required, and that's just to get you started.

Though not without its challenges, I have found this type of research to be energizing work. It is a process of connecting with others, learning from their experience, and knowing something of value can be shared that might in the future contribute to another's well-being. Others have shown me how the approach can be applied in their own field of endeavor, and this chapter explores a few of these ideas, not as the sole applications for the approach but as possibilities.

POTENTIALS AND POSSIBILITIES

When a writer crafts a story based on life experience, at its best, the story will serve to connect the reader with the person whose experience is being told, allowing a deeper appreciation of what it must have been like and what it felt like for that person at that time and in that circumstance.

Rather than taking a more traditional, distanced voice often found in many academic journals and scholarly reports, gateway brings this sense of connection to research. Gateway narratives make interview data accessible by allowing the researcher to reduce volumes of data and then step out of the way, opening the door so that study participants can come to life and speak directly to the invisible audience of readers.

Rather than reducing the data into coded themes for computer analysis or identifying key elements to condense and embed in a summary, a gateway study maintains the richness of the original expression, a goal that is quite compatible with the tenets of both oral history and educational criticism. Stories evoke complex emotions and permit the reader to connect to the experience of others on many levels of understanding. Paraphrasing or summarizing would diminish the impact of the experience, but preserving the original expressions and sequencing the interviews into excerpted narratives allow data buried deep within the transcripts to recreate the experience and bring a vicarious appreciation of what it must have been like.

These basic tenets of the approach give it the potential for disclosing the very real effects of situations, programs, actions, decisions, and events on the individuals involved. The model can be used singly or in combination with other methodologies to investigate the meaning-making process and to consider the significance that individuals ascribe to their experience. When used in combination with other methodologies, the approach can create a synergy that expands their utility and broadens the potential for presentation and dissemination. Techniques for creating excerpted narratives, for example, can contribute to oral history research by providing a means to share interview transcripts and storied conversations with those who may lack access to archives and oral history databases. Embracing the narrator's viewpoint and transforming the narrator check into an opportunity to consider what has been learned can inform advocacy. The use of interviews in educational criticism can be expanded through the practical strategies and conceptual basis for interviewing. In keeping with the arts-based, literary style of presentation, excerpted narratives also offer a model for the reduction of observation data. Data presentation for an ethnography or phenomenology may be achieved through excerpted narrative and narrator check. Program evaluation could employ gateway practices to demonstrate program impacts and results. A mixed method approach combining gateway strategies with quantitative methodologies or studies of statistical databases offers the potential for making numbers come to life, so that the human effect of trends and cause–effect ratios, for example, can be made real. One researcher has advocated the use of excerpted narratives in a bully-proofing curriculum. Another has proposed using video recording of interviews packaged with excerpted narratives for advocacy to the legislature, and another has suggested similar use in professional development. The possibilities are limited only by the imagination.

One application of selected gateway techniques has involved the adaptation of the method of data reduction and preservation of voice in combination with Latina/o critical theory, "to examine social constructions of Latina/o parental involvement and track the myriad ways that race, ethnicity, and hierarchical power play out in an urban Latina/o community" (Prosperi, 2007, p. 5). Prosperi found the approach's dedication to preserving voice especially appropriate for his study, since it allowed him to capture the parents' voice in Spanish with a poetic intensity. "Lat/Crit holds dear that the *word* must be told and the *world* understood by people who come from colonized lands in the Americas, without translations by colonizers" (Tonso & Prosperi, 2008, p. 10). Thus, Prosperi preserved the linguistic character and cadence of the parents' Spanish language, making it possible for non-English speakers to move from the margins of research and be represented in ways that illuminate their humanity and the reality of their experience. The following excerpt demonstrates his strategy for preserving the voice of speakers by providing a translation to supplement instead of supplant the voice. In this way, he maintains the integrity of their expression, allowing his narrators to see their own words presented in the research while opening the door to understanding for those of us not fluent in Spanish:

La vida en Mejico

Life in Mexico

Sin . . .

. . . de orígenes humildes . . .

un pueblito . . . un ranchito chiquitito

sin luz, sin agua, sin nada.

Yo nada más hasta al grado seis . . .

y fue mucho. . . . los mayores

sin escuela . . .

. . . Without . . .

. . . of humble origins

a little town . . . a very small town

without light, without water, without anything

I only (had schooling) until grade 6 . . .

and that (schooling) was a lot . . . for the older ones went

without schooling . . .

(Prosperi, 2007, p. 113)

While not the intent of the gateway approach, it is possible that the process offers participants a positive outcome in that they have an opportunity to reflect on their own experiences as objectified stories with an external presence, instead of as an internally housed set of memories. Seeing one's story in print, especially for those from groups traditionally underrepresented in research, is a validating experience. The narrator check is an essential factor in this process. Participants have told me that reading their experiences in a sequenced sort of story gives them great satisfaction. One told me that when she glanced at her transcript and saw how she jumped around (as is common in spoken communication), she was concerned that she sounded incoherent. When she read her excerpted narrative, she said that she was pleased to see that it communicated exactly what she really felt and had been rambling around trying to say. She also saw the themes and patterns that had previously escaped her notice.

One of my favorite stories of the power of the narrator check comes from a woman who read her narrative and then said, "This is beautiful. I completely agree with it, but I didn't say that. It's just too beautiful and I don't talk like that." When I pointed out the exact passage in her transcript, she stared in disbelief. She simply did not realize how exquisitely she had communicated her thoughts. In this case, the narrator gained an appreciation not only for her experience but also for her ability to express it.

A participant of a different gateway study said that when he saw his fragmented recollections as a "whole," with events sequenced chronologically into a story, he was able to consider all that he had done and all the challenges that he had faced. Previously, he had not had the luxury of perspective, but now it all made sense. He was able to share a deeper understanding and pride for what this experience meant to him.

Counselors who have viewed the model have suggested that its design is consistent with a therapeutic approach and that certain elements offer tools that may be useful in their applied practice. In *telling* one's story, a narrator gives voice to an experience and receives the validation of having someone listen. The benefits of sharing one's experiences with a willing and interested listener are well known, and putting words to an experience is a key component of counseling that seeks to support individuals in reclaiming their voice by telling of troubling events and feelings. "Telling one's story is an empowering experience that potentially restores a sense of continuity and wholeness" (Richman, 2006, p. 639).

Similarly, the *reading* of one's own words, organized into a coherent presentation constructed from the meanderings and out-of-sequence recollections that characterize memory, allows an individual to "hear" the meaning that he or she has taken from an experience. An overall story of life events begins to take shape. Challenges and accomplishments that might have been overlooked or disregarded become clear. Life choices and their consequences are revealed, and patterns that were previously invisible begin to emerge. This *hearing* of one's own voice allows a greater understanding of personal experience and its meaning or significance.

Gateway inquiry can help traverse limits of experience, connecting the researcher, the narrator, and the reader by bringing the qualities of a situation or event into vibrant clarity so that all can achieve a deeper understanding. However, the gateway approach is not intended to be therapy, and while therapists might employ these practices, researchers should be alert to any temptation to view it as such or to offer advice or counsel. A responsible researcher needs to be vigilant against interviewing from a perspective of offering treatment and, even though there may be a potential for positive outcomes for participants, must not confuse research with therapy.

Researching Issues of a Sensitive Nature

Throughout this text, I have given examples of fairly routine matters that could be investigated with the gateway approach. I have intentionally avoided offering a preponderance of examples for using the practice as a way to learn from traumatic situations because, quite truthfully, I have feared that gateway would be pigeonholed as having utility only for investigating disaster. This is clearly not a *trauma methodology*, but it does have utility for investigations into cases of trauma and circumstances of personal challenge.

Documenting stories of life-changing or life-challenging events requires stepping into a world that can be troubling and strange. In shedding light on painful experience, the demands can be overwhelming, whether the study focuses on the trauma of exposure to violence; weather-related disaster within a community; or discrimination and persecution because of race, gender, culture, disability, or economic disadvantage. In some cases, there can be great risk, both to the narrator and to the interviewer as well. In the chance that your research question addresses matters of particular sensitivity, I offer the following observations from my own research into the impacts of trauma. These are simply my personal opinions and are not meant to be viewed as all-inclusive or authoritative advice.

Know yourself. A key piece of doing any research into an issue that might be termed *sensitive* is that you come to terms with your own responses to it. That doesn't mean that you need to become coldly detached or devoid of all feeling. It simply means that you must be clear of your own psychological relationship to the topic before heading off to work with others. It means coming to terms with the depth of personal revelation that may be shared and developing a frame of mind that supports the other in bearing witness.

A strategy that I have found helpful is to be interviewed for my own history related to the specific event or to a circumstance similar to what I am researching. This exercise can uncover layers of response and meaning that might otherwise lay buried until some time when, out in the field, they may arise in response to a narrator's story. Being the hearer of highly personal or troubling stories carries an enormous responsibility to the narrator. The significance of receiving such stories is not to be minimized, because to ask someone to tell his or her story, and then back away, emotionally victimizes the narrator. To me, this constitutes a breach of moral responsibility. If you do not feel confident in hearing difficult testimony, please consider changing the focus of your investigation.

Understand the unique demands of the setting. To research an experience of a troubling nature, you need to acquire an awareness of the details of the event and the impact on those who lived it so that you will be able to record the story from the perspective of others. If the event involves trauma or violation of any kind, remember the powerful effect of the trauma membrane (Chapter 1) that creates a sense of separation, a feeling that others won't be able to understand the depth of the experience. In this situation, it is advisable to spend time within the community, becoming familiar with the environment and culture, and perhaps linking with an agency or nonprofit organization that can help you cross the boundary of experience so you can hear the stories of those inside the event.

If you are researching a situation involving tragedy or victimization, you need a working knowledge of the human response to trauma, not only as a way of preparing psychologically for the interview but also because some characteristics of trauma response have particular significance to anyone attempting to collect life stories in that arena. When a tragedy strikes, the challenge that victims and their families face is a difficult one. "Traumatic events frequently leave their survivors with memories that seem frozen in time—often their stories are set apart . . . by language use, tense, and form of address—as though these memories are isolated in memory" (McMahan & Rogers, 1994, p. 32). Words seem to take on new meaning, yet they fail to convey the significance that the individual may want to communicate. People now must accommodate into their worldview an experience that they had not assumed to be likely. "Basic assumptions about themselves and their world, built over years of experience, have been shattered and a new assumptive world—one that incorporates the experience of victimization—must be rebuilt" (Janoff-Bulman, 1985, p. 31).

Interviewers entering this scene need a basic understanding of the human response to extreme distress so that they do not expect participants to behave in ways that might be anticipated in a "normal" setting. If you plan to conduct research of a sensitive nature, prepare yourself by learning about the situation and common responses that you might encounter. Speak with someone outside of your research sample, someone with a similar background or exposure, who is willing to help you prepare for the interviews.

Be careful in scheduling the interviews. Surviving a trauma is not a single event, with life quickly restored to preexisting normalcy. It requires adapting to a new awareness about life and possibly a new sense of *normal.* This process is not a linear one that all individuals complete at the same rate or in the same sequence. Scheduling interviews with a narrator before that person is ready to process and share troubling memories places that individual in jeopardy. In addition, you need to be alert to complicating factors that may emerge, because the aftermath of trauma is often characterized by a series of related events or developments that cause additional disruption.

Talking about troubling or sensitive issues can be emotionally and physically draining. Scheduling interviews without giving your narrator and yourself time to rebound from these demands places you both at risk. The amount of time to allow between each interview varies with the intensity and with the individual, so it is important for you to observe and respect your narrators' needs as well as your own.

Be alert for possible negative responses. Persons setting out to collect stories of a sensitive or painful nature are well advised to critically assess whether this task poses the potential of retraumatization. To prompt a narrator into recollections that he or she is trying to avoid increases risk. If your interviews will address issues of a sensitive nature, consider asking your narrators to tell a close friend or family member about their participation in your research. It is also advisable to develop, in advance, a plan for dealing with emotional responses or discomfort. Work with each participant to develop a specific action plan, telling you what they would do if troubling feelings resurface, what resources they would seek, and whether they have insurance coverage should therapy be needed. Agree on a simple hand motion that narrators can use if they need to stop for a moment to collect their thoughts, regain composure, or terminate the interview. This nonverbal signal gives narrators control over the situation, without requiring that they explain or describe their reason for wanting to stop. In addition, you need to be alert to observe any complicating factors that may emerge and then pause the interview to allow the narrator to take a break. Refrain from counseling or giving advice but be aware of the possibility that you may need to gently end the interview. If troubling memories are shared during an interview, a phone call to check in with the narrator the following day may be in order.

Take care of yourself after the interview. Listening to stories of a sensitive nature can be difficult, so you need to plan strategies for dealing with your own stress. Vicarious, or secondary, traumatization is not unheard of among those who work in this arena. It may be helpful to have someone to talk with and debrief your interviews. Be certain, however, not to disclose identities or reveal any confidences that you must safeguard. Consult with your academic advisor or a member of your research committee regarding this issue, and if you need to speak with a counselor about matters that trouble you, do not hesitate to get help.

Recognize the power of listening. In recording and studying memories related to a circumstance that has been experienced by many others, you may find that your narrators' memories do not match the views of others and that their personal interpretations may be lost in the larger societal view. By giving an individual the opportunity to voice personal memories, a listener helps that person, in essence, validate his or her own experience. "Listening is such a simple act. It requires us to be present, and that takes practice, but we don't have to do anything else. . . . Whatever life we have experienced, if we can tell our story to someone who listens, we find it easier to deal with our circumstances" (Wheatley, 2002, para. 3).

Research is not intended to be therapy, and you need to self-monitor and refrain from offering advice or well-intended suggestions. Be vigilant against interviewing from a perspective of judgment and refrain from confusing research with treatment. However, it is possible for ethical, sensitive, and informed research that respects the authority narrators have over their experiences and their opinions, to produce positive outcomes for the narrator as well as for the research itself.

In an in-depth interview, the narrator is guided in reviewing life experience. In shaping one's story, a narrator gives voice to an experience and receives the validation of having someone listen. This process supports individuals in reclaiming their voice, their own understanding. Similarly, the reading of one's own words, organized into an excerpted narrative, allows individuals to reflect on the meaning that they personally take from their own experience. At the same time, many take pride in knowing that they have transformed their memories into valued lessons that could be used to help others.

In the Columbine study, parents had initially agreed to participate because they saw it as a way of helping to inform other communities about the impacts of a school shooting and what could be done to promote recovery. One parent declared that he would allow me to interview him because he saw it as a way of serving others. He quickly added that, while I had his permission to share his story in my dissertation, he did not want to see it in "Barnes and Noble Bookstore for public distribution." He was willing to cross the trauma membrane to share with crisis responders, therapists, and educators in ways that could help, but he did not want to "go public" with such a private experience. By the time I had finished the research and this father saw his story in print, he told me that it needed to be shared–"in Barnes and Noble"– so that others could understand what happens after a devastating tragedy and what needs to be done to help. He had come to terms with what he had shared and felt pride in having made a contribution to the well-being of others. The research provided a gateway for others to understand the event, but it also provided a gateway for those inside to reconnect with a larger world.

The Future of Gateway

Cognitive meaning and perception are only a small part of landscape of human experience. The gateway model charts a path to access a community of experience by learning from narrators about the social, emotional, physical, psychological, spiritual, and other qualities of the experience. Its attention to the voices of those who lived an event or circumstance contributes the potential to share a more holistic understanding of the complexities of life situations so that those who make decisions or plan action related to such situations may be better informed. It can be a way of linking worlds.

By attending to the complex human dynamics that are often invisible in statistically based research, it is possible to learn a great many things. It is my hope that the gateway approach might contribute to this expanding awareness, so that people are able to connect with others whose experiences have been shared and whose situation may be appreciated and understood at least a little more deeply. Making research matter means transforming it from an academic exercise and putting it to task so that our inquiry has meaning.

A FINAL WORD

This book has introduced my approach to connecting the world of research and the world of experience. For the Columbine study, I cobbled together strategies and improvised techniques so that I could investigate a painful situation, learn lessons that would help others, and do so in a way that would preserve the integrity of the narrators' perceptions and emotions without imposing my own. The unanticipated result was a cohesive, responsive approach to in-depth interviewing, a narrator-centered model for research that helps traverse the boundaries of experience, connecting the researcher, the narrator, and the reader with principled transparency. While particularly useful for investigating a community-wide tragedy, it has utility for learning from other situations as well, whether an incident of historical merit, an event of social significance, or simply interesting occurrences within the range of human experience.

Each of us understands life a little differently. Our past shapes how we live our present and how we envision our future. But we can learn from each other, adjusting our awareness so we can appreciate the experience and understanding of others. In this way, we can gain a deeper understanding of ourselves. If you choose to build your own gateways, I would love to hear of your work. When there are enough investigations in this vein, I would like to see a published series of these studies, so that research—yes, even dissertation research—could be put to use in helping to build a larger community of understanding that connects us all.

Enjoy your journey.

REFERENCES

Barone, T. (2007). A return to the gold standard? Questioning the future of narrative construction as educational research. *Qualitative Inquiry, 13*(4), 454–470.

Carlile, V. L. S. (1993). *An educational criticism of an Austrian hauptschule.* Doctoral dissertation, University of Texas at Austin.

Chase, S. (2005). Narrative inquiry: Multiple lenses, approaches, voices. In N. K. Denzin & Y. S. Lincoln (Eds.), *The Sage handbook of qualitative research* (3rd ed., pp. 651–679). Thousand Oaks, CA: Sage.

CITI data confirms success of outreach and growth. (2008, February). *CITI Newsletter, 2*(1), 1.

Citro, C., Ilgen, D., & Marrett, C. (Eds.). (2003). *Protecting participants and facilitating social and behavioral research.* Washington, DC: National Academies Press.

Clark, M. M. (2002). The September 11, 2001, Oral History Narrative and Memory Project: A first report. *The Journal of American History, 89*(2), 569–580.

Cleary, R. E. (1987). The impact of IRBs on political science research. *IRB: Ethics and human research, 9*(3), 6–10. Retrieved July 12, 2008, from http://www.jstor.org/stable/3563744

Creswell, J. (2007). *Qualitative inquiry and research design: Choosing among five approaches* (2nd ed.). Thousand Oaks, CA: Sage.

Denzin, N. K., & Lincoln, Y. S. (Eds.). (2005). *The Sage handbook of qualitative research* (3rd ed.). Thousand Oaks, CA: Sage.

DiPalma, G. I. (2007). How did you do that? Analysis of the gateway process. Unpublished manuscript.

Eisner, E. W. (1998). *The enlightened eye: Qualitative inquiry and the enhancement of educational practice.* Upper Saddle River, NJ: Merrill.

Eisner, E. W. (2002). *The educational imagination* (3rd ed.). Upper Saddle River, NJ: Merrill/Prentice Hall.

Elmfeldt, J. (1997). *Lasningens roster. Om litteratur, genus och lararskap* [Alternative title: *Voices of reading. On literature, gender, and teaching*]. Doctoral dissertation, Lunds Universitet, Sweden.

Felman, S., & Laub, D. (1992). *Testimony: Crisis of witnessing in literature, psychoanalysis, and history.* New York: Routledge.

Fitch, K. L. (2005). Difficult interactions between IRBs and investigators: Applications and solutions. *Journal of Applied Communication Research, 33*(3), 269–276.

Fontana, A., & Frey, J. H. (2005). The interview: From neutral stance to political involvement. In N. K. Denzin & Y. S. Lincoln (Eds.), *The Sage handbook of qualitative research* (3rd ed., pp. 695–727). Thousand Oaks, CA: Sage.

Frisch, M. (1990). *A shared authority: Essays on the craft and meaning of oral and public history*. Albany, NY: State University of New York Press.

Glesne, C. (1997). That rare feeling: Re-presenting research through poetic transcription. *Qualitative Inquiry, 3*(2), 202–222.

Glesne, C. (1999). *Becoming qualitative researchers: An introduction* (2nd ed.). New York: Longman.

Grele, R. J. (Ed.). (1991). *Envelopes of sound: The art of oral history* (2nd ed.). New York: Praeger.

Harris, A. K. (1991). Introduction. In R. J. Grele (Ed.), *Envelopes of sound: The art of oral history* (2nd ed., pp. 1–9). New York: Praeger.

Heinige, D. (1982). *Oral historiography*. New York: Longman.

Hesse-Biber, S. N., & Leavy, P. (2006). *The practice of qualitative research*. Thousand Oaks, CA: Sage.

Howe, K., & Eisenhart, M. (1990). Standards of qualitative (and quantitative) research: A prolegomenon. *Educational Researcher, 19*(4), 2–9.

Holyan, R. (1993). *Indian education vs. Indian schooling: An educational critique*. Doctoral dissertation, Stanford University.

Janoff-Bulman, R. (1985). The aftermath of victimization: Rebuilding shattered assumptions. In C.R. Figley (Ed.), *Trauma and its wake: The study and treatment of post-traumatic stress disorder* (pp. 15–35). New York: Brunner-Mazel.

Kuzmic, J. (2006). Dr. Carolyn Mears: QRSIG outstanding dissertation award winner for 2006, *QRSIG: AERA Qualitative Research Special Interest, IX*. Retrieved April 4, 2008, from http://www.aera.net/uploadedFiles/SIGs/Qualitative_Research/Newsletters/QRSIGSummer2006_PDF.pdf

Landrum, R. E. (1993). Sensitivity of implicit memory to input processing and the Zeigarnik effect. *Journal of General Psychology, 120*(2), 91–98.

Lichtman, M. (2006). *Qualitative research in education: A user's guide*. Thousand Oaks, CA: Sage.

Lincoln, Y. S. (2005). Institutional review boards and methodological conservatism: The challenge to and from phenomenological paradigms. In N. K. Denzin & Y. S. Lincoln (Eds.), *The Sage handbook of qualitative research* (3rd ed., pp. 165–181). Thousand Oaks, CA: Sage.

Lincoln, Y. S., & Guba, E. G. (1985). *Narrative inquiry*. Newbury Park, CA: Sage.

Lindy, J. D. (1985). Trauma membrane and other clinical concepts derived from psychotherapeutic work with survivors of natural disasters. *Psychiatric Annals, 15*(3), 153–160.

Lindy, J. D., Grace, M., & Green, B. (1981). Survivors: Outreach to a reluctant population. *American Journal of Orthopsychiatry, 51*, 468–478.

Lummis, T. (2006). Structure and validity in oral evidence. In R. Perks & A. Thomson (Eds.), *The oral history reader* (2nd ed., pp. 255–260). London: Routledge.

McMahan, E. M., & Rogers, K. L. (Eds.). (1994). *Interactive oral history interviewing*. Hillsdale, NJ: Lawrence Erlbaum.

Marshall, P. A. (2003). Human subjects protections: Institutional review boards, and cultural anthropological research. *Anthropological Quarterly, 76*(2), 269–285.

Mears, C. L. (2005). *Experiences of Columbine parents: Finding a way to tomorrow*. Doctoral dissertation, University of Denver.

Mears, C. L. (2008). A Columbine study: Giving voice, hearing meaning. *The Oral History Review, 35*(2), 159–175.

Mears, C. L. (2008). Gateways to understanding: A model for exploring and discerning meaning from experience. *International Journal of Qualitative Studies in Education, 21*(4), 407–425.

Miles, M., & Huberman, A. M. (1994). *Qualitative data analysis* (2nd ed.). Thousand Oaks, CA: Sage.

Mishler, E. G. (1991). *Research interviewing: Context and narrative*. Cambridge, MA: Harvard University Press.

Morgan, B. (2007). *A description of the implementation of the Technology-Assisted Language Learning (TALL) system into two English language learning classes at a private religious school in northern Mexico*. Doctoral dissertation, Utah State University.

Moroye, C. (2007). *Greening our future: The practices of ecologically minded teachers*. Doctoral dissertation, University of Denver.

Morrissey, C. (1987).The two-sentence format as an interviewing technique in oral history fieldwork. *Oral History Review, 15*, 43–53.

Munroe, E. A. (1997). *Enhancing the passing moments: An educational criticism of family visits to an early childhood science exhibition*. Doctoral dissertation, University of Calgary, Canada.

National Commission for the Protection of Human Subjects of Biomedical and Behavioral Research. (1979). *The Belmont report: Ethical principles and guidelines for the protection of human subjects of research*. Washington, DC: Government Printing Office. Retrieved May 24, 2008, from http://www.hhs.gov/ohrp/humansubjects/guidance/belmont.htm

Norquay, N. (1999). Identity and forgetting. *Oral History Review, 26*(1), 1–22.

Nuremberg Code (1949). In *Trials of war criminals before the Nuremberg military tribunals under Control Council Law No. 10,* (Vol. 2, pp. 181–182). Washington, DC: U.S. Government Printing Office. Retrieved June 2, 2008, from http://ohsr.od.nih.gov/guidelines/nuremberg.html

Onwuegbuzie, A. J., & Leech, N. L. (2007). Validity and qualitative research: An oxymoron? *Quality & Quantity, 41*, 233–249.

Oral History Association. (2000). *Oral history evaluation guidelines*. Retrieved March 18, 2008, from http://alpha.dickinson.edu/organizations/oha/pub_eg.html

Oral History Association. (2003). *Institutional review boards and human subjects research*. Retrieved June 9, 2008, from http://alpha.dickinson.edu/organizations/oha/org_irb.html

Oral History Society. (2007). *Getting started: What is oral history?* Retrieved February 16, 2008, from http://www.ohs.org.uk/index.php

Patai, D. (1988). *Brazilian women speak: Contemporary life stories.* New Brunswick, NJ: Rutgers University Press.

Patton, M. V. (1990). *Qualitative evaluation and research methods* (2nd ed.). Newbury Park, CA: Sage.

Peshkin, A. (1988). Virtuous subjectivity: In the participant-observer's I's. In D. N. Berg & K. K. Smith (Eds.), *The self in social inquiry: Researching methods* (pp. 267–281). Newbury Park, CA: Sage.

Pipher, M. (2006). *Writing to change the world.* New York: Riverhead Books.

Plummer, K. (2005). Critical humanism and queer theory: Living with tensions. In N. K. Denzin & Y. S. Lincoln (Eds.), *The Sage handbook of qualitative research* (3rd ed., pp. 357–373). Thousand Oaks, CA: Sage.

Portelli, A. (1991). *The death of Luigi Trastulli and other stories: Form and meaning in oral history.* Albany, NY: State University of New York Press.

Portelli, A. (1997). *The battle of Valle of Guilla: Oral history and the art of dialogue.* Madison, WI: University of Wisconsin Press.

Prosperi, J. D. H. (2007). *Mexican immigrant parents and their involvement in urban schooling: An application of Latina/o critical theory.* Doctoral dissertation, Wayne State University (Detroit).

Richardson, L. (1991). Postmodern social theory: Representational practices. *Sociological Theory, 9*(2), 173–179.

Richardson, L. (1992). The consequences of poetic representation: Writing the other, rewriting the self. In C. Ellis & M. G. Flaherty (Eds.), *Investigating subjectivity: Research on lived experience* (pp. 125–140). Thousand Oaks, CA: Sage.

Richardson, L. (1993). Poetics, dramatics, and transgressive validity: The case of the skipped line. *The Sociological Quarterly, 34*(4), 695–710.

Richardson, L. (1996). A sociology of responsibility. *Qualitative Sociology, 19*(4), 519–524.

Richardson, L. (2005). Writing: A method of inquiry. In N. K. Denzin & Y. S. Lincoln (Eds.), *The Sage handbook of qualitative research* (3rd ed., pp. 959–978). Thousand Oaks, CA: Sage.

Richman, S. (2006). Finding one's voice: Transforming trauma into autobiographical narrative. *Contemporary Psychoanalysis, (42)*4, 639–650.

Ritchie, D. A. (1995). *Doing oral history.* New York: Twayne Publishers.

Rubin, H. J., & Rubin, I. S. (2005). *Qualitative interviewing: The art of hearing data* (2nd ed.). Thousand Oaks, CA: Sage.

Schacter, D. L. (2001). *The seven sins of memory: How the mind forgets and remembers.* New York: Houghton Mifflin.

Schacter, D. L., Norman, K. A., & Koutstall, W. (1998). The cognitive neuroscience of constructive memory. *Annual Review of Psychology, 49,* 289–319.

Schuman, D. (1982). *Policy analysis, education, and everyday life.* Lexington, MA: D.C. Heath and Company.

Seidman, I. (2006). *Interviewing as qualitative research: A guide for researchers in education and the social sciences* (3rd ed.). New York: Teachers College Press.

Shopes, L. (2002). Making sense of oral history. *History matters: The U.S. survey course on the Web.* Retrieved March 31, 2008, from http://historymatters. gmu.edu/mse/oral/

Sparkes, A. C., Nilges, L., Swan, P., & Dowling, F. (2003). Poetic representations in sport and physical education: Insider perspectives. *Sport, Education and Society, 8*(2), 153–177.

Stiles, W. B. (1993). Quality control in qualitative research. *Clinical Psychology Review, 13*(3), 593–618.

Tedlock, D. (1991). Learning to listen: Oral history as poetry. In R. J. Grele (Ed.), *Envelopes of sound: The art of oral history* (2nd ed., pp.106–125). New York: Praeger.

Thompson, P. (2000). *The voice of the past: Oral history.* Oxford: Oxford University Press.

Tonso, K. L., & Prosperi, J. D. H. (2008, March). *En sus voces [In their voices]: Creating poetic representations for critical-theory research about Mexican-immigrant parental involvement.* Paper presented at the American Educational Research Association Annual Conference. New York, NY.

Townsend, R. B., & Belli, M. (2004). Oral History and IRBs: Caution urged as rule interpretations vary widely. *Perspectives.* Retrieved June 9, 2008, from http://www.historians.org/perspectives/issues/2004/0412/0412new4.cfm

U.S. Department of Health and Human Services. (2004). *Guidelines for the conduct of research involving human subjects at the National Institutes of Health.* Washington, DC: Government Printing Office. Retrieved June 2, 2008, from http://ohsr.od.nih.gov/guidelines/GrayBooklet82404.pdf

Uhrmacher, P. B. (1991). *Waldolf schools marching quietly unheard.* Doctoral dissertation, Stanford University.

Uhrmacher, P. B. (2001). Elliot Eisner. In J. A. Palmer (Ed.), *50 modern thinkers on education: From Piaget to the present* (pp. 247–252). London, Routledge.

Uhrmacher, P. B., & Matthews, J. (2005). *Intricate palette: Working the ideas of Elliot Eisner.* Columbus, OH: Merrill/Prentice Hall.

Weitzman, E. A. (2004). Advancing the scientific basis of qualitative research. In *National Science Foundation workshop on scientific foundations of qualitative research* (pp. 145–148). Retrieved April 4, 2008, from http://www.nsf.gov/pubs/2004/nsf04219/start.htm

Wheatley, M. (2002, January). Listening as healing. *Shambhala Sun.* Retrieved September 24, 2008, from http://www.shambhalasun.com/index.php?option=com_content&task=view&id=2247&Itemid=0

Yow, V. R. (1994). *Recording oral history: A practical guide for social scientists.* Thousand Oaks, CA: Sage.

Yow, V. R. (1997). Do I like them too much? Effects of the oral history interview on the interviewer and vice-versa. *The Oral History Review, 24*(1), 55–77.

Yow, V. R. (2005). *Recording oral history: A guide for the humanities and social sciences* (2nd ed.). Walnut Creek, CA: AltaMira.

Appendix A

Resources and Technology Tools

The work of in-depth interviewing takes time—time to acquire background knowledge, time to consider and recruit narrators, time to prepare thoughtful interview questions, and time to transcribe, manage, and analyze interview data. To help you with the logistical challenges you will face, there are many electronic resources and digital tools available. There is also practical advice to be found in books and articles written by researchers and practitioners who have penned their suggestions for in-depth interviewing and oral history work. This section describes a few of many useful research tools currently available.

Many of the resources described here I use in my own work; other promising resources have been suggested by colleagues. This is not intended to be an exhaustive or complete list. Please do not consider this an endorsement to buy any of these items nor a promise that any of them will be exactly right for you. You need to discover that for yourself, but I want to make you aware of some of the options. Regarding the resources that are available through the Internet, be aware that some of the URLs may have changed by the time you are reading this text, but if you type the name of the organization or the technology tool into your browser, you should be able to locate the resource.

The resources and tools described in this appendix are arranged in the following order:

Appendix A-1: Web-based resources
Appendix A-2: Recording devices
Appendix A-3: Voice recognition software (VRS)
Appendix A-4: Reference search tools
Appendix A-5: Bibliographic management tools
Appendix A-6: Recommended reading

APPENDIX A-1: WEB-BASED RESOURCES

Many organizations and institutions offer support and practical guidelines for the in-depth interview researcher. Among the selections below, you will find FREE access to resources providing advice on using technology in research, ethical standards for research and publication, copyright regulations, online oral history workshops, listservs of researchers and practitioners to connect with and ask advice on designing your study, and a variety of other useful tools.

Resources on in-depth interviewing for research

American Historical Association
http://www.historians.org/index.cfm
A nonprofit organization for the promotion of historical studies, the collection and preservation of historical documents and artifacts, and the dissemination of historical research, providing links to resources and advice for graduate students and researchers. Article from online publication *Perspectives*, "Taking a Byte Out of the Archives: Making Technology Work for You" by Kirklin Bateman, Sheila Brennan, Douglas Mudd, and Paula Petrik, is available for FREE at http://www.historians.org/Perspectives/Issues/2005/0501/0501arc1.cfm

American Psychological Association
http://apa.org/
Professional organization representing those in the field of psychology in the United States. *The Publication Manual of the American Psychological Association* describes the editorial style and formats that are used by researchers in many disciplines, including social sciences and education. For information on the publication style manual and the companion software (*APA-Style Helper*), consult http://apastyle.apa.org/ Also from APA, FREE downloadable guidelines and standards for ethical practice and research, including issues related to informed consent to research, institutional approval, recording, reporting, and publication are available at http://www.apa.org/ethics/code2002.html

Baylor University Institute for Oral History
http://www.baylor.edu/Oral_History/
FREE access to a comprehensive online workshop on in-depth oral history interviewing, with downloadable manual covering such topics as planning a project, legal documents, choosing and using equipment, selecting interviewees, and much more.

History Matters
http://historymatters.gmu.edu/mse/oral/
FREE download of *Making Sense of Oral History* by Linda Shopes (2002), an overview of oral history and the ways it is used by historians, tips for interviewing, interpreting an interview, and other resources.

H-Net Humanities and Social Sciences Online
http://www.h-net.org/
An interdisciplinary organization of scholars and teachers dedicated to developing educational potential of the Internet; publishes peer-reviewed essays, multimedia materials, and discussion points for the purpose of facilitating the free exchange of academic ideas and scholarly resources; sponsors over 100 FREE electronic, interactive newsletters and listservs edited by scholars in North America, South America, Europe, Africa, and the Pacific.

Indiana University Center for the Study of History and Memory
http://www.indiana.edu/~cshm/index.html
A grant-funded research center dedicated to building upon work in the field of oral history while broadening the range of its research projects to address the many ways that people remember, represent, and use the past in public and private life. FREE online guide to *Oral History Interviewing Techniques: How to Organize and Conduct Oral History Interviews* by Barbara Truesdell.

International Oral History Association
http://www.ioha.fgv.br/
A worldwide network of oral history scholars, researchers, and practitioners, with a Web site that makes available FREE articles and information of interest to oral historians and in-depth interview researchers.

Oral History Association
http://www.oralhistory.org/
An organization dedicated to gathering, preserving, and interpreting the voices and memories of people, communities, and past events. Web site offers a social network for making connections and sharing interests. (The OHA's *Evaluation Guidelines* and a variety of articles on oral history are available for FREE through this site.)

Oral History Association listserv
http://www.h-net.org/~oralhist/
FREE listserv, for those with an interest in oral history and in-depth interviewing, offered through the *H-Net for Humanities and Social*

Sciences Online; provides a forum for discussion of current issues, advice from practitioners and researchers, extensive bibliographies, and up-to-date, practical guidance on purchasing and using technology tools for in-depth interviewing. Discussion is archived for ongoing access.

Oral History Society
http://www.ohs.org.uk/
FREE practical information; "Advice" link leads to *Practical Advice: Getting Started—What Is Oral History?* with recommendations for purchasing recorders, planning and scheduling, how to interview, and other practical matters.

Qualitative Research listserv—QUALRS-L
http://www.listserv.uga.edu/cgi-bin/wa?SUBED1=qualrs-1&A=1
An electronic discussion group for those interested in qualitative research.

Smithsonian Institution
http://www.folklife.si.edu/explore/educator_resources.html
FREE download *Smithsonian Folklore and Oral History Interviewing Guide* by Marjorie Hunt, a manual that presents guidelines developed by the Smithsonian over the years for collecting oral history and conducting in-depth interviews, sample information and release forms, and suggestions for presenting and preserving findings.

Stories Matter
http://storytelling.concordia.ca/storiesmatter/
A FREE open source software built for oral historians by The Centre for Oral History and Digital Storytelling to allow archiving of digital video and audio materials, enabling the annotation, analysis, and evaluation of materials in their collections. In addition to an offline version, the software will have an online version that will facilitate sharing and collaboration in the discipline.

Texas Historical Commission
http://www.thc.state.tx.us/publications/guidelines/OralHistory.pdf
FREE download of *Fundamentals of Oral History*, a manual that provides guidelines and suggestions for designing and completing oral history projects.

University of Leicester
http://www.le.ac.uk/emoha/howtointerview/index.html
FREE access to practical advice on a wide variety of considerations for interviewing as research, including guidance on recording devices (downloadable text files as well as sound and video demonstrations).

U.S. Copyright Office
http://www.copyright.gov/
FREE information on copyrights, fair use of copyrighted material, what copyright protection means, and how to register a work.

Vermont Folklife Center
http://www.vermontfolklifecenter.org/archive/res_audioequip.htm
FREE download *of Digital Audio Field Recording Equipment Guide* by Andy Kolovos, a guide to selecting digital audio recording equipment for use in conducting ethnographic fieldwork, with advice on digital products by brand and type.

Resources on standards for professional research conduct

Code of Ethics, American Anthropological Association (1998).
   http://www.aaanet.org/committees/ethics/ethicscode.pdf
Ethical Standards, American Sociological Association (rev. 2005).
   http://www.asanet.org/cs/root/leftnav/ethics/code_of_ethics_
   standards
Oral History Evaluation Guidelines, Oral History Association (rev. 2000).
   http://alpha.dickinson.edu/organizations/oha/pub_eg.html
Standards for Reporting on Empirical Social Science Research in AERA Publications, American Educational Research Association (2006).
   http://aera.net/uploadedFiles/Opportunities/Standardsfor
   ReportingEmpiricalSocialScience_PDF.pdf
Statement of Standards of Professional Conduct of the American Historical Association, American Historical Association (rev. 2005).
   http://www.historians.org/PUBS/Free/ProfessionalStandards.cfm

Information on human subjects protection

Historians and Institutional Review Boards: A (Not So) Brief Bibliography by Linda Shopes (2005) is available through the Oral History Association at http://alpha.dickinson.edu/organizations/oha/pdf/org_irb_bibliography.pdf This annotated bibliography provides an extensive list of publications related to policy, regulations, reports, commentary, and criticism and is an excellent place to start a consideration of the implications of working with the IRB for nonmedical research.

National Institutes of Health, Office of Human Subjects Research at http://ohsr.od.nih.gov/ provides links to regulations and guidelines for the ethical conduct of research that involves human subjects. Among these documents are the following:

- *The Nuremberg Code* at http://ohsr.od.nih.gov/guidelines/nuremberg. html
- *The Belmont Report* at http://ohsr.od.nih.gov/guidelines/belmont. html
- *Title 45 (Public Welfare) Code of Federal Regulations, Part 46 (Protection of Human Subjects)* at http://ohsr.od.nih.gov/guide lines/45cfr46.html
- *Guidelines for Conduct of Research Involving Human Subjects at NIH* at http://ohsr.od.nih.gov/guidelines/GrayBooklet82404. pdf

Office of Human Research Protections/US Department of Health & Human Services has responsibility for implementing federal regulations related to protection of human research subjects. OHRP information at http://www.hhs.gov/ohrp/ includes links to Title 45 Part 46, *The Belmont Report*, IRB compliance oversight requirements, as well as additional information related to the regulations and their implications for researchers.

APPENDIX A-2: RECORDING DEVICES

To complete first-rate interview research, you need the best recording equipment you can buy or borrow. Unless you are able to record your narrator's conversation at a high enough quality that it can be heard clearly and transcribed accurately, your time spent at the interview has been wasted.

Your choice of recording equipment essentially falls within two categories, analog or digital. In general, analog devices rely on a physical tape, reel-to-reel or cassette, while digital recorders store data in a digital format on a disk drive, memory card, or other medium within the device. Analog recorders have long held sway as the recorder of choice among interview researchers, but dramatic advances in digital technology have made it clear that the advantages of "going digital" are just too many to ignore.

It sometimes seems that advancements in technology occur almost from minute to minute, and truthfully any guidelines or recommendations about brands or functionality would be quickly out of date. When you are ready to begin your research, you can find the most current information and rating of devices and technology tools by consulting the Web sites and listservs cited in the Web resource section of this appendix. Many of these sites provide a forum for researchers to update and comment on what is currently available. Also check with colleagues and friends to ask their advice. If you are unable to purchase equipment, ask around to see if you might borrow good recording devices and perhaps microphones from your university or perhaps from an individual who has had experience in recording and transcribing in-depth interviews.

As previously mentioned, I use two recorders for all of my interviewing. When I began this work, I used one analog and one digital, but now I have switched to using two digital recorders. Digital devices enable you to download a sound file directly onto your computer. This can help you manage and store your data. Whichever type you choose, be certain that the device is comfortable to work with, that you are able to see the tape counter or digital display, that it has a powerful enough microphone to pick up conversational speech, and that the speaker or "output" sound system can be easily heard. Before making a purchase, check the type of battery that is required and the anticipated battery life while recording.

For some studies, a video recorder might be useful. My advice on selecting a recorder for video interviewing is the same as for digital: Ask someone who has recently completed research using this medium about their experience. Video for data collection and for presentation can be a powerful way to complete your research; however, you need skill in editing and communications design to make the most of the medium. Also, if you are considering video recording using a digital device, please be especially vigilant in protecting video files. The ease with which it is possible to video-stream such a file to the Internet makes it even more imperative that the highest level of safeguards are in place and that your narrator has agreed to video-record with full awareness of the risk/benefit factors discussed in Chapter 3.

APPENDIX A-3: VOICE RECOGNITION
SOFTWARE (VRS)

Typing by speaking has been an elusive dream of writers and researchers for a long, long time. Since people speak at an average of 125–140 words per minute but usually type at only about one-third that rate, life would certainly be simpler if this technology could be perfected. Recent developments have brought this dream to life, and it is actually quite easy to type by speaking. In fact, I have spoken many parts of this book onto the page with aid of voice recognition software (VRS).

I first turned to this amazing technology when I needed to transcribe 35–40 hours of in-depth interviews for my dissertation. I am not a skilled typist and was overwhelmed by the prospect of sitting in front of my computer for the amount of time that I knew would be required. I considered hiring someone to do the transcriptions for me, but since I was dealing with highly sensitive issues related to trauma, I would need to hire a bonded transcriber and the cost was prohibitive. (Current rates for transcription range around $100–125 per hour of recorded tape.)

As I was struggling with this dilemma, a friend recommended that I try the VRS program *Dragon NaturallySpeaking*, which I did and have been a convert ever since. I now use it not only to transcribe interviews but also to write just about anything I want, including parts of this text. It is useful for almost any type of writing task; for example, when I am reading an article that I want to make note of, I simply open my *EndNote* file (see Appendix A-4), and then read the paragraph that I want to quote. The VRS program types it out as I read and I don't need to manually key it in. VRS can be used for most tasks that require keyboarding or word processing, including e-mail.

For transcription work, VRS has been my salvation. I simply listen to the audio recording of each narrator and then repeat the words into the microphone. The software does the rest, producing a transcript in any format that I choose, complete with punctuation and paragraphing. By using a digital recorder I am able to slow down the playing speed slightly, so that I can basically listen and then echo what I hear. I love being able to bracket nonverbals as I go along and insert notes to myself or comments that I need to follow up on.

While I am happy to be able to speak more and type less, some writers are less convinced because there are undoubtedly errors to correct. VRS programs seem to work better with some voices than it does for others. For programs that produce a 98% accuracy rate, that's only two errors per 100 words, and that's far more accurate than I'm able to type manually. It does take some getting used to, though, and the decision to try it should be based on your individual style and preferences as well as the amount of typing that you need to do.

Another factor for you to consider is computer operability. At present VRS technology requires 2-4GB RAM and approximately 4 GB free hard disk space. Be sure to check that the computer you will be using has sufficient capacity to operate and store the large digital files that will be produced.

In choosing a VRS package, you need to consider your operating system, system requirements, and the software applications you normally use. The following are among the most highly regarded options at this time.

Dragon NaturallySpeaking
http://www.nuance.com/
VRS for Windows and Apple allows dictation, editing, and formatting in such programs as Microsoft Word, Microsoft Outlook Express, Microsoft Internet Explorer, and AOL.

MacSpeech Dictate
http://store.apple.com/us/product/TR284LL/A
VRS for Intel-based Mac, written specifically for the Mac, with Dragon Speech recognition capabilities at its core.

IBM Viavoice
http://www.nuance.com/viavoice/pro/
VRS for computers with Linux operating systems.

Appendix A-4: Reference Search Tools

Many of the selected research tools described in this section are available only through subscription, and as a result must be accessed through a library or research institution. I have not provided URLs for these resources. However, you should be able to access these search tools while using computers at a research library. Some database search tools, as noted, are available at no charge. For those, I have provided the current URL.

Academic Search Premier
Research capabilities to locate articles, full-text eBooks, and other resources through institutional library services, full-text access for over 4,000 general and scholarly publications.

Academic OneFile
Source for full-text, peer-reviewed articles from over 12,000 journals and other reference publications covering topics from humanities, science, technology, and medicine.

America's Historical Newspapers
Database for access to more than 1,000 U.S. historical newspapers published during the past 400 years, including digitized versions of primary source material.

Directory of Open Access Journals (DOAJ)
http://www.doaj.org/
FREE research service covers full-text, quality-controlled scientific and scholarly journals, including access to over 3,600 journals in the directory.

ebrary
Resource for full-text books, reports, and other content; requires downloading *ebrary Reader.*

EBSCOhost
Online database for access to resources from tens of thousands of institutions worldwide, indexing, abstracts and full-text articles, archival coverage for university libraries, reference databases, online journals, books, and other services.

Expanded Academic ASAP
Access to abstracts and selected full-text material from scholarly journals, news magazines, and newspapers.

Google Scholar
http://scholar.google.com/
FREE search tool to access scholarly literature including peer-reviewed papers, theses, books, abstracts, and articles from academic publishers, professional societies, preprint repositories, universities, and other scholarly organizations.

informaworld
http://www.informaworld.com
A one-stop site hosting journals, eBooks, abstract databases, and reference works published by Taylor & Francis, Routledge, Psychology Press, and Informa Healthcare.

ISI Web of Knowledge
An integrated research environment for multidisciplinary and specialized content and the tools for access, analysis, and management of research information from more than 140,000 full-text Web documents.

LexisNexis Academic
Database for access to full-text news, business, and legal publications from over 6,000 sources, using a variety of flexible search options.

LexisNexis Congressional
Access for full coverage of U.S. federal legislative activity from 1789 to present.

National Academies Press
www.nap.edu
FREE access to full-text reports issued by the National Academy of Sciences, the National Academy of Engineering, the Institute of Medicine, and the National Research Council, all operating under a charter granted by the U.S. Congress.

Proquest
Database for access to research more than 11,000 titles for periodicals, dissertations, and digital newspaper archives research, 8,000 of which provide full-text access to scholarship from libraries and private collections around the world.

PsycARTICLES
Database containing more than 25,000 searchable full-text articles from journals published by the American Psychological Association and allied organizations.

PsycINFO
Database for access to over 1,500 journals, dissertations, book chapters, books, technical reports, and other documents in the fields of psychology, education, medicine, social science, and organizational behavior.

Questia
http://www.questia.com/Index.jsp
Access to search online library of books and articles in journals, newspapers, and magazines in the humanities and social sciences; provides search, note-taking, and writing tools to help in locating relevant information, quote and cite correctly, and create properly formatted footnotes and bibliographies automatically.

WorldCat
Database of bibliographic records for books, serials (journals), and all other formats.

APPENDIX A-5: BIBLIOGRAPHIC MANAGEMENT TOOLS

The following software programs and online applications have been developed to facilitate research across the methodologies. They offer tools for managing references, abstracting, note taking, developing bibliographies, and, in some cases, Internet searches. Social bookmarking sites allow you to store, organize, and share bookmarks of Web pages that you have found helpful in your research. Since these bookmarks are stored on the Web, you are able to access your saved information and resources even when you are not working on your own computer. As noted, many of these data management tools are available at no charge.

academia.edu
http://www.academia.edu/
FREE Web resource to connect university-affiliated academics globally, displayed in a "tree format" arranged by their university and department; academic networking to connect researchers of shared interests through Web page and scholarly publications.

Bookends
http://www.sonnysoftware.com/
A reference manager for Macintosh providing assistance with bibliography, reference, management tasks, as well as Internet searches and pdf downloads.

CiteULike
http://www.citeulike.org/
FREE service for organizing, storing, and sharing academic papers; a social bookmarking site.

Connotea
http://www.connotea.org/
FREE, no download required, a social bookmarking and citation service allows saving of references complete with URL, link to the webpage, pdf, and other sources, with online tutorial.

delicious
http://delicious.com/
FREE service for organizing, storing, and sharing academic papers; a social bookmarking site.

EndNote
http://www.endnote.com/
Software for Macintosh and Windows operating systems, assists in completing reviews of the literature and development of bibliographies by

providing templates for note taking, organizing references, formatting bibliography, locating full-text articles, and preparing Word documents. FREE trial download available. Along with *EndNote,* other reference tools, including *EndNote Web, ProCite,* and *Reference Manager,* are all made by Thomson Reuters, an information technology corporation. To compare functionality, check their Web site http://scientific.thom sonreuters.com/rs/

Papyrus Bibliography System and Knowledge Manager
http://www.researchsoftwaredesign.com/
FREE download to store and link to references, notes, images; locate and import bibliographic references; format documents; available in Macintosh and DOS/Windows formats.

Scribe
http://chnm.gmu.edu/tools/scribe/
FREE download, a cross-platform note-taking program to manage research notes, bibliographic information, quotes, published and archival sources, digital images, outlines, timelines, and glossary entries; made available from the Center for History and News Media.

Zotero
http://www.zotero.com
FREE Firefox extension for collecting, managing, and citing sources; does not work with Internet Explorer.

APPENDIX A-6: RECOMMENDED READING

The following books and print materials provide a resource for you to extend your thinking about issues of ethical research practice, guidelines for conducting in-depth interviewing, the tradition of oral history, and qualitative inquiry.

Becoming Qualitative Researchers: An Introduction (2005, 3rd ed.) by Corrine Glesne (Boston: Allyn & Bacon). A wonderfully readable overview of qualitative inquiry with attention to biographical, historical, and collaborative research practices, cross-cultural research, poetic transcription, and diverse possibilities for research.

The Death of Luigi Trastulli and Other Stories: Form and Meaning in Oral History (1991) by Alessandro Portelli (New York: State University of New York Press). A collection of articles on the method and contributions of oral history, offering examples as well as reflections on the practice of interviewing and oral history work.

Doing Oral History: A Practical Guide (2003) by Donald Ritchie (New York: Oxford University Press). Step-by-step guide to oral history, providing advice and explanations for how to interview and record stories of human experience, includes an extensive bibliography and practical tools, templates, and release forms in the appendixes.

The Enlightened Eye: Qualitative Inquiry and the Enhancement of Educational Practice (1998) by Elliot Eisner (Upper Saddle River, NJ: Merrill). Seminal work describing the practice of educational connoisseurship and criticism, its origins, its functions, and its design.

Envelopes of Sound: The Art of Oral History (1991, 2nd ed.) edited by Ronald J. Grele (New York: Praeger). A collection of articles in conversation on the theory and meaning of the practice of oral history, memory, and the significance of oral history as a method for collecting stories of experience.

Interviewing as Qualitative Research: A Guide for Researchers in Education and the Social Sciences (2006, 3rd ed.) by Irving Seidman (New York: Teachers College Press). Helpful resource with step-by-step guidance and the principles of interviewing for phenomenology that are useful to a wide range of interviewing applications.

Intricate Palette: Working the Ideas of Elliot Eisner (2005) by P. Bruce Uhrmacher and Jonathan Matthews (Columbus, OH: Merrill Prentice Hall). A reexamination of Elliot Eisner's writings, providing insightful analysis and evaluation of his work and thinking.

Learning From Strangers: The Art and Method of Qualitative Interview Studies (1994) by Robert Weiss (New York: The Free Press). A how-to-do-it book that offers insights into effective interviewing and the risks and opportunities of this type of research.

The Life Story Interview (1998) by Robert Atkinson (Thousand Oaks, CA: Sage). Brief volume with information on preparing and conducting life story interviews, transcription, analysis, interpretation, criteria for judging quality, and benefits of participating in the research.

Loss of the Assumptive World: A Theory of Traumatic Loss (2002) edited by Jeffery Kauffman (New York: Brunner-Routledge). A collection of articles on the reconstruction of meaning and worldview that is necessitated by traumatic loss; valuable preparation for those interviewing survivors of disaster and victimization.

Narrating Our Pasts: The Social Construction of Oral History (1992) by Elizabeth Tonkin (Cambridge: Cambridge University Press). A volume that considers the construction and interpretation of oral history, providing examples of studies in memory, narration, and oral tradition from different countries.

Navigating Life Review: Interviews with Survivors of Trauma by Mark Klempner, in *The Oral History Reader* (2006, pp. 198-210, 2nd ed.), edited by Robert Perks and Alistair Thomson (London: Routledge). Excellent background and precautions for interviewing trauma survivors.

Oral History and the Law (2002) by John Neuenschwander (Carslisle, PA: Oral History Association). Important discussion by an author who is both an oral historian and a judge, of legal considerations of oral history research, with examples from legal cases and their implications.

Oral History and Public Memories (2008) edited by Paula Hamilton and Linda Shopes (Philadelphia: Temple University Press). A volume of articles from a variety of research environments, a rich account of oral history and memory studies at work.

The Oral History Reader (2006, 2nd ed.) edited by Robert Perks and Alistair Thomson (London: Routledge). International anthology of classic articles on theory and practice of oral history; includes selections on methods and relationship, interviewing regarding traumatic memories, reconciliation politics, memory and interpretation, digital and technology resources, dissemination, and other topics.

Preparing Literature Reviews: Qualitative and Quantitative Approaches (2008, 3rd ed.) by M. Ling Pan (Glendale, CA: Pyrczak). A how-to-do-it manual for reviewing the literature and preparing literature reviews to support your research, emphasizing selecting a topic, locating the research literature, evaluating sources, and synthesizing literature into a cohesive review.

Protecting Participants and Facilitating Social and Behavioral Research (2003) edited by Constance Citro, Daniel Ilgen, and Cora Marrett (Washington, DC: National Academies Press). Available from http://www.nap.edu/ Overview of the history of regulations governing human subjects research in social sciences, the work of the IRB, and essential precautions to ensure protection of research participants.

Qualitative Data Analysis: An Expanded Sourcebook (1994, 2nd ed.) by Matthew Miles and Michael Huberman (Thousand Oaks, CA: Sage). Valuable tool for managing, organizing, and processing qualitative data.

Qualitative Interviewing: The Art of Hearing Data (2004, 2nd ed.) by Herbert Rubin and Irene Rubin (Thousand Oaks, CA: Sage). Practical guide to open-ended in-depth interviewing, with guidance and examples to prepare the reader for managing the complexities of designing and carrying out interview research.

Recording Oral History: A Guide for the Humanities and Social Sciences (2005, 2nd ed.) by Valerie Raleigh Yow (Walnut Creek, CA: AltaMira Press). Comprehensive text on the many aspects of oral history including the design and conduct of the in-depth interview, oral history and memory, legalities and ethics, interpersonal relations, varieties of oral history projects, analysis and interpretation, sample release forms, and more.

Research Interviewing: Context and Narrative (1991) by Elliot G. Mishler (Boston: Harvard University Press). Examination of the process and the significance of interviewing for social and behavioral research.

A Shared Authority: Essays on the Craft and Meaning of Oral and Public History (1990) by Michael Frisch (New York: State University of New York Press). A collection of essays about oral history and case studies to demonstrate oral and public history; considers oral history scholarship, memory, responsibility of researcher to interviewee, recording, presentation, quality, and more.

Shattered Assumptions: Towards a New Psychology of Trauma (1992) by Ronnie Janoff-Bulman (New York: The Free Press). A solid introduction to the trauma response and the way people come to terms with trauma and loss.

The Two-Sentence Format as an Interviewing Technique in Oral History Fieldwork, *Oral History Review,* (Spring 1987, *15,* pp. 43–53) by Charles Morrissey. Informative discussion of the structure of questioning and the importance of context for in-depth interviewing.

The Voice of the Past: Oral History (1978) by Paul Thompson (Oxford: Oxford University Press). A consideration of the theory of oral history and the technical processes involved; traces oral history through its own past to the present, discussing reliability of oral sources; considers memory, narrative approaches, technology, and the social function of historical writing.

Writing Literature Reviews: A Guide for Students of the Social and Behavioral Sciences (2006, 3rd ed.) by Jose L. Galvan (Glendale, CA: Pyrczak). Step-by-step instructions and helpful examples for developing a review of research literature.

Writing Up Qualitative Research (2001, 2nd ed.) by Harry Wolcott (Thousand Oaks, CA: Sage). Conversational approach to advice and counsel on completing dissertations and writing for research.

APPENDIX B

FORMS AND ORGANIZERS

In this appendix, you will find samples of forms and organizational templates that may be helpful to you in your research investigation. Be sure to check the requirements of the institutional review board (IRB) overseeing your research before assuming that the sample informed consent (Appendix B-1) meets their criteria.

Appendix B-1: Sample informed consent form
Appendix B-2: Checklist for gateway research
Appendix B-3: Sample interview guide
Appendix B-4: Sample interview summary form
Appendix B-5: Data analysis organizer

APPENDIX B-1: SAMPLE INFORMED CONSENT FORM

You are invited to participate in a research study of the experiences of individuals who were placed in the foster care system in Blank County. This research is being conducted to fulfill the dissertation requirement for a doctorate in Education at the University of
_____.

The Research

For this research, I am asking that you agree to participate in three interviews, 60–90 minutes each, to be conducted in a location of your choice. I will ask you to talk about your experiences while you were in the foster care system and the years following. Each interview will be tape-recorded, and you will have an opportunity to review the transcripts of your recordings to make corrections. After we have completed all three interviews, I will share with you the narrative that I have constructed from your interview transcripts and ask you to check the accuracy and completeness of the representation of your experience. You will have an opportunity to respond to my observations.

Risks

The interviews are designed to minimize any emotional or psychological discomfort to you. However, discussing your experiences might cause you unpleasant feelings, such as sadness, anger, and anxiety, or might trigger troubling memories. If at any time you feel uncomfortable or overwhelmed, you are encouraged to request a break or to terminate the interview. Likewise, if I observe that the interview appears to trouble you, I will suggest that we pause or terminate the session.

Before we begin the first interview, I will ask you to help identify resources to which you would turn should you experience difficulty as a result of these interviews. In addition to personal resources you may have available to you, please be aware that the Blank County Health Department currently provides support services on sliding-scale fee basis to residents of the county. As a resident of the county, you may be eligible for these services. Their phone number is 555-555-5555.

Benefits

The benefits of participating in this study are that you will have an opportunity to contribute your experience and understanding of what it is like to be placed in foster care. In reflecting on and sharing your experiences, you will be contributing to the awareness of the complexities of being a child in this type of program so that future programs can be designed and conducted in such a way as to be of maximum benefit to those who are enrolled. You will help to inform service providers and policy makers who must plan resources and support for children in the future.

Confidentiality

Every effort will be made to ensure that the information you share with me will remain confidential. My dissertation co-chairs will have access to your interview data, but your name will not be used in my dissertation, and all identifying information will be deleted or abridged in order to protect your identity. This assurance of confidentiality extends to all members of your family. In addition, your participation will not be shared with others in this study.

By signing this form, you acknowledge that you understand there are two exceptions to the promise of confidentiality. If information is revealed that concerns homicide, suicide, or child abuse and neglect, I am required by law to report this information to the proper authorities. In addition, if any information in this study is subject to a court order or lawful subpoena, the University might not be able to avoid compliance with the order or subpoena.

Special Considerations

Please know that your participation in this study is entirely voluntary. You may, at any time, decline to answer any question without having to qualify your reasons for doing so. You may, at any time, request a break, terminate the session, or remove yourself from this study, without any penalty or loss of benefit, and without having to qualify your reasons for doing so. You may withdraw from this investigation with full confidence that any information you have shared will not be included in the study. You will be given a copy of your interview transcripts for your records. If you decide to remain in this study, you will also receive a copy of the research results.

Whom to Contact

If you have any questions about this study, please call me at 555-5111, Dr. Smith, my dissertation co-chair at the University at 555-5222, or Dr. Jones, also a dissertation co-chair at the University, at 555-5333. Also, if you have concerns or complaints about how you were treated during the study, please call Dr. Jane, Chair, Institutional Review Board for the Protection of Human Subjects at 555-5444, or Dr. John, Office of Sponsored Programs 555-5666, or write to either at the Office of Sponsored Programs, University Campus, P.O. Box 555.

If you agree to these statements and conditions and you agree to participate in this study, please sign below.

I have read and understand the foregoing description of this research project. I have asked for and received a satisfactory explanation of any language that I did not fully understand. I agree to participate in this study, and understand that I may withdraw my consent at any time. I grant the use of my interview for this dissertation and for any publications or presentations that are based on this dissertation research. I have received a copy of this consent form.

Signature: _____ Date: _____

Please print:

Name: _____

Address: _____

_____

Phone: _____ E-mail: _____

_____ I give my consent to be audiotaped.

_____ I do not give my consent to be audiotaped.

Signature: _____ Date: _____

Thank you so much for your interest in this study.

APPENDIX B-2: CHECKLIST FOR GATEWAY RESEARCH

Check when completed	Steps for gateway research	Chapters
	Conceptualize clear purpose for the research	2, 5
	Frame research questions that can be answered through in-depth interviews	2, 5
	Develop antecedent knowledge to allow informed listening, interpretation, and analysis	2, 4, 5, 8
	Identify population to be focus of the inquiry	3, 5
	Set criteria for selecting narrators and develop strategy for access	4, 5, 6
	Design participant safeguards	3, 8
	Propose research and get consensus of research committee (dissertation advisors) as to viability of investigation	5
	Apply for and get clearance from sponsoring institution (IRB, human subjects review board, etc.)	3, 5
	Conduct series of in-depth interviews, usually three sessions per narrator, depending on complexity of research questions	5, 6
	Transcribe interviews	6, 7
	Review/interpret transcriptions to discern each narrator's meaning relevant to research questions and to identify patterns or themes in their reflections (vertical, in-depth interpretation for each narrator independently)	7

	Prepare excerpted narrative for each narrator	7
	Meet with narrator to confirm accuracy, adequacy, and completeness of story as representing narrator's intended meaning and understanding of event/ circumstance being researched (narrator check)	7
	Revise, correct narratives as needed	7
	Analyze all narratives to identify common patterns/themes, discern benefiting and inhibiting factors, and develop recommendations (horizontal analysis across all narratives)	7
	If appropriate, develop mosaic to present combined experience represented in the research sample	7
	Present findings in ways that disclose human experience, significance, and meanings taken from the setting or circumstance	7

APPENDIX B-3: SAMPLE INTERVIEW GUIDE

Primary research question: How does participation in an experiential literacy program affect the students who are enrolled?
Sub-questions: What are the benefits and what are the impediments to participation? What are the outcomes for students? What recommendations would students make for improving the program?

FIRST INTERVIEW
Introduce the project and tell what to expect

Explain purpose for the research, what I am attempting to learn, and how research will be used and shared. Tell a little about my interest in the project. Tell how I got their name and why I selected them to participate. Explain the interview process, why it is being recorded, what to expect in each session, etc.

Informed consent

Review in detail the Informed Consent Form and ask them to sign a copy. Give them a copy of the form for their records.

Open questions to help frame discussion to follow

Ask, *I'm interested in learning about the experiential literacy program and would like to hear of your experiences in the program at East Shore.*
From resulting narratives look for effects and ask follow-up questions related to academic growth, socialization, family, self-esteem, career decisions, and other topics that emerge that are related to the research question.
Ask about key points from narrative, for example, *You mentioned that you didn't get along with some of the other students in the program and even felt threatened. Tell me a little about that.*
Ask, *Tell me what you've been doing since you left high school.* [This is connected to outcome.]
As time allows, ask for examples or stories, feelings about or reactions to the experience, and changes that participation brought.
At the end of the interview, explain that next time you will explore some of these areas more deeply. Ask your narrator to make a note of anything that comes up in the time between the interviews that might be of interest.

SECOND INTERVIEW

Give narrator an opportunity to return to what he/she talked about in the first session and explore experience in greater depth.

Ask, *Was there anything that we talked about last time that was particularly meaningful to you?*

Encourage additional disclosure and stories by guiding narrator to topics that may have been introduced but not fully considered in first interview.

Return to the key points concerning <u>academic growth</u>, <u>socialization</u>, <u>family</u>, <u>self-esteem</u>, <u>career decisions</u>, etc. for deeper reflection. Address questions or inconsistencies from transcript from session one. Get clarification, confirm your understanding, and ask about topics that weren't covered (see Interview Summary Form). Let narrator know what you might like for him/her to be thinking about for the next interview.

THIRD INTERVIEW

Address any topics on summary form that have not been fully explored. This session also allows for reflection.

Ask, *If you could go back and give yourself advice about participating in the program, what would you say?*

Ask, *What would you tell teachers and administrators that they need to think about when planning this type of program?*

Ask for metaphors to describe the experience, *What was it like?*

Ask, *What did you think I might ask you in these interviews?* If you haven't covered that topic, do so now.

Ask, *Since your interviews have covered a lot of territory, not all of it can be included in my report. What would you be disappointed to see left out?*

NARRATOR CHECK SESSION

Ask narrator to review the narrative you created for accuracy and completeness: *This is how I interpreted what you've shared with me in the previous sessions. Did I get it right? What needs to be fixed? Clarified? Deleted?*

Ask, *Do you see anything in the narrative that surprises you?*

Ask, *Did the narrative I constructed using your words remind you of anything you'd like to add?*

Ask, *How do you feel about being interviewed about experiences in the program?*

Don't forget to express your appreciation for their help!

APPENDIX B-4: SAMPLE INTERVIEW SUMMARY FORM

Interview Summary Form
Analysis of literacy program

Narrator Ref. Code _____
   Interview number: _____ Interview date: _____
   Today's date: _____

1. Main themes or issues that became apparent during this interview
2. Observations
3. Information that relates to the research questions (note page of transcript)
4. Particularly salient stories on the reverse side of this form.

Question	Pg.	Information	Comment
Experience in program			
Effects while in HS [outcomes]			
Effects after HS [outcomes]			
Challenges faced as student			
Resources and supports			
Benefits			
Impediments			
Recommen-dations			
Other			

APPENDIX B-5: DATA ANALYSIS ORGANIZER

Narrators Themes or patterns	Narrator 1	Narrator 2	Narrator 3	Narrator 4
Theme 1				
Theme 2				
Theme 3				
Theme 4				
Theme 5				

Appendix C

Sample Narratives

The selections in this appendix demonstrate various stages of data display using techniques from poetic transcription to create excerpted narratives. The first example provides excerpts from a transcript and demonstrates the stages of development of a narrative for a mid-career doctoral student. You might work with this narrative to practice identifying key elements that you would include in an excerpted narrative and seeing what you consider to be the *only* words. The second example is a finalized display with the profile of the narrator that introduces a complete excerpted narrative for a Columbine mother. The final example demonstrates the use of excerpted narrative to present the collective voices of all participants in the Columbine study.

Appendix C-1: Sample transcript and narrative
Appendix C-2: Profile and complete narrative
Appendix C-3: Columbine Mosaic

Appendix C-1: Sample Transcript and Narrative

The excerpts below are from transcripts of interviews with Bill, a mid-career marketing executive who was completing his doctoral degree in business administration. **The first question I asked was,** *Tell me a little about yourself and your life as a grad student. What brought you to the point that you are in your career and in your life?*

Bill's response

Well, to begin with I grew up in Pittsburgh, Pennsylvania, and went to Catholic grade school and high school and my first undergraduate degree was in business administration at _____ University. And then I decided in my last class of my undergraduate degree that I wanted to major in international business and so I went to my professor and he said that the school you'd have to go to is called _____ in Arizona. It's the best school for international business. So I applied to the school and promptly got rejected because my grade point average wasn't high enough. And I called up the admissions director, and said if I got a second bachelor's degree could I be considered based on my second degree's grade point average as opposed to an average. And he said Yes, and I said well make sure you write that down and Dr. _____ who I remember said OK so I went off and for the next 2 years I worked full time and then I went to school full time at night for a second bachelor's degree in computer science and I completed that in 2 years and then I had a 3.6 GPA and then I was so excited and I called up the director and I was wondering if he was still there and fortunately he was and I said, Do you remember me? And he said, well let me pull out your file. So I said well I completed a second degree. He asked me what my grade point average was and I told him, and he said well congratulations you're accepted and I said well that's great. I finished in a year and a half and then I had a masters in international business.

Later in the transcript

It was only once I reached the end of the program that I realized it was the beginning. And it was only when I took that class in my last semester in international business that the switch flipped and I realized there's a whole world that I haven't seen.

Still later in the transcript

I work full time and I take one 3-credit class at a time for eight weeks and they are sequential. And then I have a one to two week break in between classes. Actually I'll receive a doctorate in business administration.

Toward the end of the first interview, I asked, *What have been some of the things that have helped you in this process in getting through your degree?*

Bill's response

I would say it's a personal resilience that I just refuse to quit, which others might view it as a counter-dependency. If someone tells you that you can't do something then I will show them that I can. It's my immediate response is that it's my refuse-to-quit attitude that serves me. A lot of that comes from my upbringing. It also comes from sports where you're taught not to quit so no matter how bad it gets you don't quit. And that's the way I feel. There is many times that I've felt exhausted—but I just didn't quit. There were times when the assignments just seemed insurmountable in the amount of work and research and I would just try to break it down into something that I could do this week and just take it a week at a time. Just a step at a time and say okay "this project, the literature review, you've got eight weeks to finish this for your first draft." This is huge, it's gathering resources coming up with a draft—it's just huge and I would set little goals. If I could find 10 resources this week I'll set that as the goal. And then toward the end you always have to go faster because they'll find out that I might be behind a little bit but it worked out. In the end I took off three days of work that week that I have the final draft due so that I could write from eight in the morning until 2 a.m. and for five days I wrote from eight in the morning until 2 a.m. And at the end of that week I finished because I wasn't going to quit.

In the second interview, I asked, *How do you feel about this thing called the dissertation?*

Bill's response

I'm the type of person that works well under pressure so if I know that I've got a certain time frame I will meet that deadline. I can honestly say in the entire program up to this point I've never missed a deadline.

In working with the full transcripts that are excerpted above, I noted a pattern in Bill's approach to school and completing his degree. Bill is successful in his work, in his education, and in life itself, because of his resiliency and his refusal to back away from a challenge or leave a job undone. I read through Bill's transcripts and excerpted passages that spoke to this pattern of resilience and dedication to task completion.

I grew up in Pittsburgh, Pennsylvania and
went to Catholic grade school and high school
my first undergraduate degree was in business administration at _____
    University.
then I decided in my last class of my undergraduate degree that I
    wanted to major in international business
I went to my professor and he said that the school you'd have to go to
    is called _____ in Arizona.
It's the best school for international business.
I applied to the school and promptly got rejected because my grade
    point average wasn't high enough
I called up the admissions director, and said if I got a second
    bachelor's degree could I be considered based on my second
    degree's grade point average as opposed to an average.
And he said Yes,
so I went off and for the next 2 years I worked full time and then
    I went to school full time at night for a second bachelor's degree
    in computer science
I completed that in 2 years and then I had a 3.6 GPA
then I was so excited and I called up the director and I was wondering
    if he was still there and fortunately he was and I said, Do you
    remember me?
He asked me what my grade point average was and I told him, and he
    said well congratulations you're accepted—
It was only once I reached the end of the program that I realized it
    was the beginning
And it was only when I took that class in my last semester in
    international business that the switch flipped and I realized there's
    a whole world that I haven't seen.
I work full time
I take one 3-credit class at a time for eight weeks
I'll receive a doctorate in business administration.
I would say it's a personal resilience that I just refuse to quit
It's my immediate response is that a refuse to quit attitude that
    serves me.
A lot of that comes from my upbringing.
It also comes from sports where you're taught not to quit so no
    matter how bad it gets you don't quit.
And that's the way I feel.
There is many times that I've felt exhausted—but I just didn't
quit.
There were times when the assignments just seemed insurmountable
    in the amount of work and research and
I would just try to break it down into something that I could do this
    week
just take it a week at a time.

Just a step at a time and say okay.
I would set little goals.
I could write from eight in the morning until 2 a.m. and for five days
wrote from eight in the morning until 2 a.m.
At the end I finished because I wasn't going to quit.
I'm the type of person that works well under pressure
I can honestly say in the entire program up to this point I've never
   missed a deadline.

Next, I further reduced the narrative into the barest of expressions, finding what I considered to be the *only* words and condensing the presentation to its simplest form without sounding cryptic. I rearranged a few of the statements so that it had a sequenced flow, yet preserved the meaning as Bill had spoken it. Bill's narrative, following, reveals the importance of this work ethic and resilience, traits that placed high in his recommendations to other students pursuing advanced degrees.

I grew up in Pittsburgh
went to Catholic school
decided in my last class of my undergraduate degree
I wanted to major in international business.

My professor said,
Best school for international business is in Arizona.
I applied
got rejected
grade point average wasn't high enough.
I called the admissions director,
I said, If I get a second bachelor's
could I be considered on second degree's grade point average?
For the next 2 years
worked full time
went to school full time at night
second bachelor's—computer science
3.6 GPA
Called up the director
Congratulations you're accepted.

Only when I took that class
in my last semester
in international business
that the switch flipped
The end was the beginning
I realized there's a whole world that I haven't seen.
Finished in a year and a half,

Had a masters in international business
A doctorate in business administration

I just refuse to quit
A refuse-to-quit attitude
Personal resilience
A lot of that comes from my upbringing.
It also comes from sports—
No matter how bad it gets you don't quit.
I've felt exhausted—but I just didn't quit.

Assignments seemed insurmountable
Just a step at a time
Set little goals
Write from eight in the morning until 2 a.m.
In the entire program I never missed a deadline.

I finished
I wasn't going to quit.

Appendix C-2: Profile and Complete Narrative

Note: To introduce the narrative for each parent in the Columbine study, I prepared a brief sketch or profile, almost as an artist would create a line drawing to reveal contours but not the details of a scene. The details come later, in the parent's narrative, told in his or her own voice, enriched through its presentation in a poem-like structure. Each narrative is constructed from direct quotations excerpted from the parent's interview transcripts and arranged into a narrative sequence. My guidelines for creating these narratives were (1) include only the passages that speak to my research questions; (2) rearrange the sequence of the excerpts if that improves clarity and effectiveness; and (3) only change words if re-sequencing the narrative necessitates changes in verb form or pronoun/noun reference for clarity. In an effort to protect the confidentiality of study participants and their families, all names were changed. Details that might have provided clues to their identity were deleted. Each participant confirmed that the narrative is accurate and represents his or her actual experiences, responses, and insights.

Lillian

With quiet grace, Lillian sits, awaiting my first question. "Just let me know what you need to know," she says, eager to help make this study a success. Lillian's tastefully decorated living room, with subdued lighting and soft sounds of classical music gently wrapping around us, is the setting as we began our discussion of the deadliest of rampage shootings to date. In her story, Lillian remarks on the many paradoxes and ironies of the brutal attack in the presumed safety of a library, in a school noted for its excellence. Our relaxed conversations about violence and bloodshed—while sitting in this charming and tranquil home—present a similar undercurrent of incongruity.

Lillian is a single mother with two children—Jeff, age 26, and Jenny, age 22, both Columbine graduates. Lillian is a native Coloradoan, a fact that she points to as having helped her in the aftermath. Family is important to Lillian, and having family and friends nearby provided much needed support. She and her husband, from whom she divorced many years ago, moved into the Columbine area when it was first being developed in the 1970s. At that time, Pierce Street, which passes in front of Columbine High School, was a dirt road, and this area of unincorporated Jefferson County was a small, quiet, middle class neighborhood with a "country atmosphere." The area has grown since then. Pierce Street is now a busy five-lane local artery, but a hint of the country feel remains, with cottonwood trees nestled along the narrow, grassy creek beds and nothing to block the view of the foothills to the west.

Lillian's two children attended Columbine area schools from kindergarten through to their graduation, and the loyalty that they feel to their neighborhood is strong. Before the shootings, Columbine was seen as a safe community, where families worked hard to ensure that their children had good lives and good futures. The April 20th assault not only threatened their assumptions about personal safety, but also challenged the identity of their high school and their community.

Lillian's daughter Jenny was a junior at Columbine at the time of the attack. Jenny wasn't in the building when it occurred, but like other Columbine students, her world was deeply shaken just the same. Jeff, a junior in college that year, was not even in town when the attack took place, yet his world was afflicted too. And when the world of her children is shaken, so, too, is a mother's.

Lillian's intelligence, her quiet strength, her love of family, and her commitment to church and community are reflected in her narrative. When asked for an action plan, should she become troubled by memories of this tragic event, her answer was a quick one, "I have a great support system," she said. "My friends, my family, my church."

Lillian offers the perspective of a single mother who joined with her children, her family, her friends, and her community, in grieving the violent loss of life in a school and in acknowledging the vulnerability that faces us all. Here, in her own words, is Lillian's story.

April 20th . . .

It was lunchtime. I was coming down Wadsworth.
All of a sudden,
Three cop cars going so fast—
About 6 feet apart.
It scared me to death—I thought,
Oh my gosh, there's been a wreck.
It has to do with high school kids.
You know how crazy the kids are in cars.
That's always been my biggest fear.

I decided to go past the school,
See what was going on.
The street was all blocked off.

I saw kids running.
One said, There's somebody inside shooting people up.
There's people dying in there.
I just went home—
    less than a mile—
Was Jenny okay?
I got home
Her car was in the driveway.
So the only horrible minutes I had,
  I am so lucky,
Was from Leawood to home.

Jenny was here.
    Other girls too,
        Jenny and her friends.
Jenny didn't open a book all through high school.
She didn't study.

They had lunch together—
      Gone to the mall to eat.
They were going to come back here to watch TV.
They had dropped a boy off at school.
He went in,
Into the library to study.
He was shot.

When I got here,
They had the TV on.
They were on the phones
      Calling their moms and dads,
Telling everyone they were okay,
Looking for the other girls in their group.

There were people at my work,
      Couldn't find their kids.
      Didn't know where they were.
I went back to work for a little bit—just to tell my boss,
I wouldn't be in the next day.
I look back and think,
That was so dumb to go back to work.

It hadn't sunk in with me that much.
I had to take care of everybody else.
    Moms were all talking
    Neighbors were all talking.
    Phones were ringing.
    Relatives calling, you know.
I wanted to take care of my mom and dad too.

That night churches all opened up,
Light of the World is a huge church.
We went there that night.

In the days following ...

Churches were doing all sorts of things for the kids.
They had group discussions for them.
One church brought in a bunch of golden retriever puppies.
Jenny just loved that.

Churches rallied to do a pretty good job.
My church didn't have resources.
    Ended up falling apart
    Our pastor had to leave.
It's huge because churches are a huge part of our lives.

Things that were bad before got exacerbated—
You know, how you can be on the edge just a little bit
And then when Columbine happened,
Those things went over the edge.

I had to take some days off.
I didn't want to leave Jenny by herself.
I went to some of the funerals, and
Some of the memorials.
Sometimes I didn't go,
Jenny went with her friends.

It's so important to go through the ritual,
You know that you can't do anything to solve it
Or to make it better right away, so
Everybody takes flowers to the park
        or go to ceremonies,
        or to churches,
        or to funerals.
You feel like you're doing something.
Once you have a ceremony or a ritual,
You can't go back to the way it was before.
You've done something.
You've taken a step,
A rite of passage.

Here it is, the kids were in the library
They're the good students,
        Studying for their tests,
How ironic, how paradoxical,
Here Jenny is, la da da da da,

And it doesn't really matter, if I get good grades or not.
I'm not going to the library.
She's out and about, but
What could be a safer place for kids than in a library?
You'd want your kids to be in a library where they're safe.
And then look at what happened.
We're all hanging by a thread.
You live with it.

Jenny's older brother Jeff was away in college.
He had the flu and was deathly ill.
I never went up there to see him.
I mean it just was the flu.
I just had to be with Jenny.
He came home that Friday night.
I think being home with the family,
Actually seeing that Jenny was okay
That she was alive,
He was much better.
This was so traumatic for him.
Sometimes I think this affected him worse than it did Jenny,
Because he was so violated that
Someone would dare come and do this.

He knew the brother of one of the gunmen very well.
He called Jeff at school and asked, Is your sister okay?
He was crying. It just broke him up.
He was so afraid something had happened to Jenny.

At Jeff's college, as part of their student services,
He could have seen a counselor if he wanted, but
He was in a fraternity, a great network of friends.
Those guys really helped.
The following year, Jeff had a professor
Who was studying Columbine.
Jeff called Jenny to come to talk to his class about it.
I remember it clearly
Because it was the day before Greg committed suicide.

Connections . . .

During the shootings,
One of the first calls to the police was from a cell phone.
You know there were a lot of kids in the choir room
     Calling their parents.
You know it was a lifeline.
After April 20, I went out and bought Jenny a cell phone.
I think we all did, didn't we?

After that, they wouldn't turn their cell phones off at night.
They'll let anybody in the world bother them at night.
I'm saying, Jenny, turn your phone off.
She'd say, Mom what if somebody needs me in the night,
I don't get the phone call, and
Then I don't help them?

Jenny had a disease when she was younger.
I almost lost her one night in the hospital.
She almost died.
We were kinda close because of that.
I just think life is so bittersweet that I don't expect much more.
And it makes you a little paranoid about what's coming next.
What am I'm going to have to deal with next?

The shootings happened in Jenny's junior year.
It was hard on her when she had classes at Chatfield.
There was an emptiness in one of her classes.
One of her friends from that class had been killed,
And she missed her, I know she did.
She felt guilty because she was there and her friend wasn't.
Kids have a lot of guilt, but they just survive I think.

The media and outside scrutiny...

The media did a terrible job—they lied a lot.
It was always slanted,
They love to glorify the dirt.
They take things out of context.
It was just so negative all the time about the community.
I think they had a real self-righteous attitude.
They were pointing their finger, trying to make us feel dirty.
Like, What kind of community are you
To allow this to happen to these kids?
What are you doing to kids so that they feel so left out?

The people in this neighborhood,
    Work their tails off,
    Pay their taxes,
    Are good citizens,
    Do their civic duty,
    Try to do their part,
    Take good care of their kids,
Try to make sure they have good lives,
And I don't like other people trying to make me feel dirty,
Or bad.

Like, when Jenny went to college and
Met roommates from out of state,
Their parents would be curious, like,
Is Jenny going to go psycho on us?
They wondered if we're all psychological basket cases.

Jenny was on a school spirit team.
We had gone to a competition in California,
Right before April 20th.
All of the moms went and were having the best time,
Just such a fun-loving time.
The next year, after Columbine had happened,
We were competing in Florida.
Everybody was there and we were nervous.
We knew people were watching us.
And we didn't want to call attention to ourselves.
We didn't want to call attention to the girls.
We didn't want to give anybody room to say—
You know, No wonder it happened. Or, Is this why it happened?

The school . . .

The school was kind to them,
　　　　so much support.
Oh my goodness, people would help people,
If they asked for help.
People were trying.
I'm not saying they were perfect,
　　　　but they were trying.
The teachers, oh my gosh,
The teachers were rock stars.
Some of them were just awesome.
Like the AP English teacher.
She cared so desperately about the kids,
Always available to them.
Have you ever known a more gracious woman?
If the kids needed something, all they had to do was go in.

That's why I think it's so sad.
I think people don't know there's help there,
Or they don't seek help.
There are saints in this world who would give you help.
All you have to do is ask.
But there's some responsibility on yourself,
And people don't know to ask.
It is so sad that it breaks my heart.

Jenny's business teacher knew someone
Who helped them make glass balls with columbines inside,
Made them as Christmas ornaments
    Did a business plan—
    Sold ornaments—
    Made quite a bit of money—
Then took the money and
Sent it to Kosovo.

That's what it's all about.
By doing stuff like that,
You're giving back.

The art teacher was another saint.
She has a different type of personality but
A wonderful woman.
She had an artist in residence for them—Clarissa Estes—
The one who wrote *Women Who Run with Wolves*.
She would tell the art students stories to help them through this.
You see, she heals through storytelling.

Stories about if you make something good out of something that bad
Then you can redeem it somehow.
If you can't, then evil always wins.

Jenny has a lot of friends.
Her true personality is that of an art student.
She really can identify with the kids who were alternative.
But she used to say, Oh I am so marginalized.
I don't have black lipstick and pink hair, and
Everyone thinks that I can't do art,
Just because I choose to have more of a preppy look.
All kids have their judgments.

You know Jenny did the tile project
That the art teacher started.
She does have a tile in the building.

Parenting . . .

A long time ago, friends in the neighborhood
Started our own Mom's day out program for the kids,
We bonded then.
Some of us had to go through divorces and stuff
But we had these women friends.
We had a support system, and
Going through Columbine, I knew I was very lucky.
I don't know what I would've done without the other moms.

You know the sticker that said,
We are all Columbine?
I've heard a couple of kids say they didn't like it.
They'd say, You weren't there.
You don't understand.
And I'd say, No, no, no. Don't you see that's a very, very good sticker?
Because what it means is that
We know it could happen to us,
We are all vulnerable.
Because it happened to you,
And because we love you,
It happened to us too.
But they would not buy that.

The Greg thing—that's the hardest thing,
Because as a parent, I missed it.
I missed it, and this is what kills me the most.
Jenny and he were friends.
As the next spring progressed, he started dissociating himself,
Pulling away from people,
Pulling away from Jenny.
It bothered her really a lot.
Jenny would ask,
    Should I call him?
    What should I do?
I'd say, Jenny, You know what,
You can't make someone like you if they don't.
I thought it was this girl - boy thing.
Then Jeff asked her to come to talk to his class at college, and
When she came home, she said,
    Should I call Greg?
But I'm saying, You can't make someone like you if they don't,
So why don't you move on.
Yet she knew that there was something really wrong.

The next day, I got this call at work.
Jenny said, Mom, somebody has committed suicide from our
    school.
Mom, they say it's Greg.

Later, she came home.
I met her at the door.
She said, It was Greg.
She burst into tears.
I don't think I saw her cry like that ever for the Columbine thing.
She felt so guilty.
She was so distraught.

I took her to Columbine Connections.
She went two times, then
They said, You're healthy.
So she picked up and she went on.

Sometimes we're asked to endure too much.
You have to make sure you survive—
Use that to help others.

Support . . .

People sent cards and gifts to the school, and
Jenny helped with the thank you notes.
She did about 40 or 50
I was really glad she did.
There was one guy, and you know,
You could tell he was a maintenance man
Or a janitor someplace in New Jersey.
Well, they wrote back and forth.
He was an old guy and
It just delighted him to get letters.
Then she went off to college.

I think the kids got tired of it,
Tired of being constantly identified with it.
I think there was so much guilt, because they weren't hurt.
They were feeling so bad,
Like her friend got killed and she didn't.
There was just a razor thin chance that she didn't.
But doing stuff like that you're giving back.
Even though it feels like it doesn't mean a whole lot,
It does.
You're doing something.
That's desperately important.

Words of advice . . .

After April 20th, I took off of work for a little bit.
It's really hard to take off because
I always feel like I'm leaving them high and dry.
I feel a real obligation not to let my coworkers down, but
As I look back, I think I should have taken more time off for myself.
I would tell others to take off a lot more time from work.
It would help parents to grieve and heal,
I think we were all so worried about the kids,
We forgot ourselves a little bit.

The kids were in the house a lot.
I never said anything about it,
But it was hard on me.
It was hard to have people around
A challenge to not say anything.
There wasn't any downtime after work
Because of all the kids here.
As a working mom, I thought it was a time crunch.
Everything was a huge challenge for me.
But I'd rather have the kids come here,
Than to be out somewhere.
You know, out on the street somewhere.
I think it was a traumatic enough thing,
I should have forgotten about work and stayed home,
I should have taken more time off just time off to sit on the patio.

A challenging situation . . .

You know, at my church one night,
Someone said, You know,
We could look at Columbine
    As a gift to the community.
    To wake us up—
    To see what's important in life.
He said, You know, we are well-to-do down here.
We all have cars and things.
He acted like nothing bad ever happened in this community,
I thought I was going to die.
There's just no way that I could ever think that
Columbine could be a gift.

This is the part that hurts me the most about Columbine,
I know how hard people work down here and
How everyone does their part to make it a nice neighborhood.
I mean, a lot of people care.
It's bittersweet.
We have our elements, but there are a lot of good things
That happen in this neighborhood.

We talked about it being a gift but I will never buy that.
It was not a gift.
They're not thinking about the kids whose lives are ruined, or
Whose lives are taken away,
The ones who are paralyzed.
It's abhorrent that anyone could call it a gift.
I wish more than anything in the whole world that it hadn't happened.

One of the pastors said, Christ wouldn't be ministering to the
  cheerleaders
  Or the jocks.
He would be taking care of the kids
Sitting in a corner, their head down.
Well, that just infuriated Jenny.
She said, You know, I had to come back from a serious disease,
And he's telling me I don't need Christ because I'm preppy?
He's telling me that Greg didn't need help,
Because he's a basketball player?

Finding support...

  I had support, you know,
      My family
      My mom and dad,
      My sister
      My friends
      My church.
  I am really lucky.
      I had lots of support.

  But you know a neighbor said,
  She didn't have support.
  I think that she didn't go get it.
  Our community was great for support.
  It was there for the asking.
  They had Columbine Connections.
  Anybody could have had that.
  All you had to do was call.

  People across the nation sent support
      And money
      And donations
      And lots of resources.
  All the fast food places gave the kids free food.
  The grocery store down here,
  The minute it happened,
  They were giving out food and ribbons.
  I don't know how the community could have been more supportive
      than this.
  But support goes two ways.
  You have to make the effort.
  It's important,
  When you go through life to make your support system.
  Relationships are everything.
  You know that if you had your family in town, it helped.

You go to your family first.
One thing to remember is that if something happens to a loved one,
And they're someplace else,
Then you need to go wherever they are.
When a bad thing happens, you go to your family.

We bought our first house here in 1974.
Pierce Street was a dirt road.
Columbine wasn't even built yet.
I like it down here.
My family is all here—
I felt sad for those who didn't have their families around.
But, go ask for support,
People have a need for relationships.
They have a need to give,
But don't just sit there and think it's going to flow into you.
You need to let people know you need help.
The best way to get support sometimes
Is just go help somebody else.
Like Gerde, she came to the school and told about
All the things she went through in the Holocaust.
Here's this woman who has gone through so much
Who understands—
And instead of griping about, poor me,
Look what she gave back to the kids.
Look what she has done.
I can't imagine a more beautiful person.

My sister and I went to visit the Holocaust Museum after that.
If you survive you're supposed to survive.
You have to make sure you survive and use that to help others.
And the things that touch our children's lives, and helped them
Mean so much to us.
That's what I would consider a gift.

Significance and perspective . . .

If I had to tell people what happened,
I would say,
        Two boys had become evil,
        Lost their minds,
        Wanted to kill everyone.
They were very, very sick boys,
        Crossed over the edge,
        Acted as if they were possessed,
        Went into this school,
        Started this rampage.

And how lucky that for some unknown reason—
As bad as it was—
With all of these kids being injured and hurt
And even though there were 13 that died—
It could've been so much worse.

The kids were just blindsided.
Who wants to think their classmates would do such a thing?
I don't know whether to think that these kids turned to a kind of evil,
Or to pity them because they were so sick.

The tragedy is, whoever would think
That your classmates would want to shoot you?
    You know them.
    You sit next to them.
These two boys made a big deal about it, saying
We go to school
We joke with them, and
They bought all of our lies.

You know the father of the lie would be the devil.
That's just pure evil.
And they bought into the other side.
They crossed the line.
They went to the dark side.
What makes a person go to the dark side?
What causes so much rage and hate?
But you sit next to someone and
You don't even know who the enemy is.

You do suffer in this life.
I would tell people,
We're all just hanging by a thread.
Life is bittersweet and you have to take the bitter with the sweet.

You know it's such a complex thing.
I think there are lots of truths to take from it.
If you just ask the school and the teachers
And the people who held office or had titles,
You wouldn't have the full picture.
You might have 72 pieces of the puzzle
But still be missing 28 others.
When I think about the shooters' parents—
I could never condemn them because
I know that as human beings we tend to be in denial,
But there are so many signs and symptoms.
We are in denial until it's too late.
We all deny things.

So one truth would be, to be vigilant.
Being a single mom,
It just gets really hectic and you let things go that you shouldn't.
But we're human beings and we float until we can't anymore.

I wonder sometimes if the kids that have gone through this,
If maybe all of a sudden, they see what's really important to them.
And that's so different from what we teach them is important.
They see it differently, when they have come close to death.
They have a vision of their own mortality, and
Then they reassess.

If someone asked me,
When we will get over this, I'd say,
I've heard this description:
We all have a house that we live in,
In our house we have different rooms.
Columbine is one of the rooms in our house.
And it's always going to be in our house and
It's not going to go away.

It doesn't mean that we have to go in that room everyday,
But it's part of who we are now.
It's always there and how often you choose to go in—
How often you need to go in—
How often you do go in,
So be it.
Someone who doesn't have that house in their life,
Someone who doesn't have that room in their house,
Probably can't relate to it.
They just don't know.

(FROM *EXPERIENCES OF COLUMBINE PARENTS: FINDING A WAY TO TOMORROW*, pp. 76–86)

APPENDIX C-3: COLUMBINE MOSAIC

NOTE: To present a view into the collective experience of all parents who participated in the Columbine study, I crafted the following mosaic, excerpting phrases from the individual narratives so that a clearer answer to my research question regarding the experience of these parents might be shared. This narrative blends the diverse voices and responses to the shooting and reveals varied and multiple lenses to the event and the aftermath.

Columbine Mosaic

Shooting
Children slaughtered
Kids flying down the hall
Teachers got them out
Stay down, stay down!
Run!
Sniper on the roof
Moments of terror
Kids in the library
What safer place?
Blindsided

People couldn't find their kids
Lady, you can't go in there
Where is my child?
Lists on the wall
You're a victim too
Not a damn thing you can do
Helpless
A war zone
My only child
Phones ringing all night
Worried about the kids
Which kids are we going to know?

Churches opened up
Not religious
But church was a magnet
Go through the ritual
Go to memorials
Funerals
But no closure
At least you're doing something

Friends supporting each other
Overtaken with grief
God, it was raw
Emotionally nothing left
Vulnerable
Empty
Guilty for surviving
Guilty for finding my child so soon
Bittersweet
We all bought cell phones, didn't we?

Kids, just normal kids, did this.
How could they do this?
Were they sick? Evil? Needed help?
No one paying attention?
Bullying?
Mental health overlooked.
Not enough money for schools.

Media—sell the story
Make a buck
Never telling the whole truth
Pissing people off
Were you there?

Gray Line tours
Visitors taking pictures
Pictures of the kids
Wanting to go through the school
Some people wanted to help
Some proselytized
Some exploited

Kids tired—constantly identified with it
Don't call attention to themselves

Family helping each other
Friends, counselors, church
Relationships forged in trauma last a lifetime
Move outside your own pain
Deal with a whole community of pain
Move to another level of experience
A more complete person

Don't buy into the stigma
Use experience to help others

Crowded at Chatfield
Just opened the doors to Columbine kids
A fantastic thing
But treading water
Going back to Columbine, after it was cleaned up
Helped us understand.
This is where I was
This is where I hid
This is what happened to me

First day of school next year
Parents were there
Students went in
Teachers were inside
Shook their hands
So emotional
Incredible courage to go back in
Reclaim their school
It was heroic

School was kind to them
Good communication
Maybe too much
Teachers were rock stars
Busted their butts
There for the kids
Continuing to teach
Always available
Grateful to the teachers
Cared so desperately
Tile project
Storytelling
Money to Kosovo
Beat up old computers
Choir concert
Surprise Party
Extra counselors on staff
Understanding

Removed the library
Made the atrium
Beautiful
Therapeutic

Such a complex thing
Not a single event
Shootings
Suicides

Subway murders
Threats to the school
Zero tolerance
All intertwined
Sad for those without family around

Community under a microscope
Columbine equals tragedy
But shootings were not Columbine
Community helped each other
People cared
Strangers came to help
Gerde survived the Holocaust
Came to help
If you survive you were supposed to
Then help others

Some dads were, OK get over it
Don't want to talk about emotions
Suck it up
Mothers talked
Some moms were, You need therapy
Rough times around the house
On the job training
Forgot ourselves
100% attention on the kids
No time for each other
Friends out of state don't understand
Tears
Divorces
Kids hanging out
Hanging by a thread
Nightmares
Anxiety
200 kids wouldn't leave their houses
Others taking risks
No fear
Might die tomorrow anyway
Careless
Alcohol
Signs of what to look for
Is it adolescence or trauma?

My child asked
Would I be the same otherwise?
Just trying to get back to normal
What is normal?

Be vigilant
Listen
Take time with your kids
Forget about work
Lives changed
Never forget,
Just learn from it
Keep moving on

It's a chapter in my life
A room in my house
Piece of a pie
We're like damaged goods
Survivors of Hiroshima
Deep, deep gut-like worry
A hell of an experience
A tornado
A hangover
Getting punched in the stomach for weeks on end
A sad, sad time

Victims' parents—so vocal
Hurt other kids
They don't trust now
If my kid were killed I might feel different,
But don't hurt the survivors
Need a Memorial for the lost
For the living too
Those who survived
Have to live with it

Point to the stories
To the people who died
And those who survived
The injured and their recovery
A true reflection of the community

Individuals have to heal together
Bond with people who were part of it
Remember that people helped
Millions of people
Don't feel so alone with it

Feels like pain will never end
It will
Start new traditions
Seek out your friends
Don't turn away from the pain

You will have joy again
Don't let it define you
Hope
Remember what's good about the community
People holding hands
People being together
Rebuilding trust
Love in the place

If it could happen here
It could happen anywhere.

(Excerpted from *Experiences of Columbine Parents: Finding a Way to Tomorrow*, pp. 199–202)

AUTHOR INDEX

SUBJECT INDEX

33417719R00134

Printed in Great Britain
by Amazon